HEINEMANN STUDIES IN BIOLOGY Number 2

Editorial Adviser: S. A. Barnett

Cell Biology

Maurice Pallett
Pharmacology
U.C.C.

Cell Biology

by John Paul
F.R.S.E., M.B., Ch.B., Ph.D., M.R.C.P.(Ed. & Glas.), M.C.Path.
Director of Cancer Research Department,
Royal Beatson Memorial Hospital, Glasgow

SECOND EDITION

HEINEMANN EDUCATIONAL
BOOKS LTD: LONDON

Heinemann Educational Books Ltd
London Edinburgh Melbourne
Singapore Johannesburg Hong Kong
Auckland Ibadan Nairobi Toronto

SBN 435 62691 4

First published 1965
Reprinted 1966 *with additions*
Second edition 1967
Reprinted 1969

Published by Heinemann Educational Books
48 Charles Street, London WIX 8AH
Printed in Great Britain by Butler & Tanner Ltd,
Frome and London

Preface

In retrospect the past decade may turn out to have been the most dramatic period in the history of biological science. The secrets of the genetic code have been revealed, the complex structure of living molecules has been elucidated and we have all but created life in the test-tube. The background to these advances has been a ferment of research activity, the publication of enormous numbers of research reports and the emergence of the new term, molecular biology. In the face of these remarkable developments most biologists have found it impossible to keep abreast of all the new knowledge and only fragments of it have been filtering down to many undergraduates. Although perhaps unavoidable this is regrettable for much of the new information simplifies the understanding of biological principles. One particular unifying hypothesis – the cell theory – has emerged with new significance and impact as a result of these advances and this book is an attempt to present an account of the current situation in simple terms. It is primarily designed for post-graduate and senior undergraduate students in the biological sciences and therefore assumes a general acquaintance with elementary biological and chemical terminology; however, I hope that enquiring laymen and senior school pupils will not be deterred from dipping into it. The material itself cannot fail to fascinate although I may often have failed to do it justice. Where my shortcomings are particularly obvious and should not be allowed to go uncorrected I hope that informed readers will take the trouble to write and take me to task.

I have tried to outline what I consider to be the essentials of cell biology, with as little elaboration as possible, so that the reader may have an opportunity to see the whole subject in perspective.

To make up for the inevitable inadequacies I have provided a carefully selected bibliography, which ranges from relatively simple introductory articles, through books and articles providing extensive reviews, to key scientific papers reporting original findings. These are intended to serve as a bridge between this book and the very extensive scientific literature which every serious student will want to explore.

I am greatly indebted to the many individuals who have helped me in assembling the material. Mr Robin Callander's collaboration in preparing the illustrations was indispensable. His impeccable draughtsmanship is apparent in all the line drawings. Mrs Rae Fergusson's secretarial help was equally essential and her conscientiousness facilitated the task immensely. I had the benefit of useful criticism from Professor J. N. Davidson and Mr S. A. Barnett both of whom read the manuscript. Mr Barnett's meticulous editing was particularly valuable. To illustrate the text I have been exceedingly fortunate in being able to borrow illustrations from many eminent colleagues, who are individually acknowledged in the legends to the photographs. My own contribution was mainly one of collation. This is, however, very time-consuming and, when the muse has gone elsewhere for a time, not conducive to equanimity. I must, therefore, not omit to pay tribute to the patience and forbearance shown by my wife.

J. P.
Glasgow, June 1964

Note to 1966 Reprint

A number of minor revisions have been made in the text, six new plates have been added, and new references have been included in the bibliography.

Preface to Second Edition

In the two years since I wrote the first edition the study of the cell as a focal point in biology has attracted more attention than ever before. New journals have appeared, new societies have emerged and formal instruction in cell biology has been initiated in many universities. Accordingly, in preparing this edition I have had in mind the need to state clearly the problems which are of concern to cell biologists. I am sometimes asked where the boundaries of cell biology stop. There are, of course, no boundaries to cell biology any more than to any other intellectual discipline – only a focal point of intense common interest from which other interests spread, merging into the related fields of biochemistry, genetics, histology, biophysics, microbiology, protozoology and so on. The idea that scientific fields should have clearly delineated boundaries is as out of date as the mediaeval concept of the self-contained city-state. I want to make it clear, therefore, that cell biology is not restricted to the subjects I discuss in this book; they merely indicate the interests that currently stand in the centre of the field.

The preparation of a second edition has given me the opportunity to correct errors in the first, to fill in some gaps and to bring the text up to date. A new chapter has been added on interrelationships among cell structures and many new plates and figures have been included. The bibliography has been expanded to include the most significant works of the last two years.

I am grateful to those who pointed out errors in the first edition or took the trouble to write with suggestions. In the accumulation of material I have depended very much on the kindness of many colleagues who have freely given me permission to use their photographs; they are individually acknowledged in the legends. Once again I have been especially indebted to Mr Robin Callander for his great skill in preparing the drawings and to Mrs Rae Fergusson for her conscientious secretarial help. To all of these it gives me great pleasure to record my gratitude.

J. P.
Glasgow, January 1967

Contents

List of Plates

Editorial Note

Numbers in square brackets thus, [132], refer to the bibliography.

Introduction

Although everybody has an idea of what is implied by the terms 'life' and 'living' it is remarkably difficult to give them precise meanings. Indeed, some people have objected to their use because of this. As soon as we begin to think deeply about them, we encounter a multitude of problems, many unsolved and some, perhaps, insoluble. Is there a purpose in life? If so, what is it? Questions of this kind have occupied the minds of philosophers since man first began to think about himself. We have no universally acceptable answers. It would be easier to write a book on a biological subject if we had.

There is much to be said for considering life as a physical state in which matter and energy interact as in other physical states. It is a state characterized by phenomena which have to do mainly with its remarkable organizational complexity. The unique structures of living things depend on the existence of particularly large complex molecules and special supramolecular arrangements. Specialized mechanisms are required to capture and utilize the energy necessary to form them. Furthermore, perpetuation of the state depends on maintenance of this organization and hence on special replicative mechanisms.

An entirely mechanistic approach may be criticized quite justifiably on the grounds that it avoids issues of a philosophical nature. But these criticisms are outweighed by the great advantages of considering living phenomena in this way, since, if organisms can be thought of in the same terms as machines, it makes many of the problems associated with them both more tangible and more tractable. Consequently, this book is devoted to considering the cell as a biological machine.

Like machines, cells can be described in purely physical terms and many cellular processes have analogies in man-made machines and reactions. It is not too far-fetched to draw a parallel between a cell and a factory, provided we do not overlook the dramatic differences between them. Many more processes go on in a cell than in most factories, with greater efficiency and in an infinitely smaller space. *Chlamydomonas*, a unicellular organism, uses solar energy to synthesize hundreds of chemical compounds, some of which have not been made by organic chemists. The many reactions involved are automatically regulated, so that they operate in perfect harmony. Periodically, every part replicates exactly. In a population of cells experiments are constantly going on to evolve new processes which replace the old if they prove better. Yet the entire 'factory' is too small to be seen with the unaided eye.

Remarkable as the complexity of cell structure is, the most impressive feature of cellular organization is the elegance and beauty of some of the patterns which underlie it. These range from the orderly assembly of molecules in cellular architecture to the precise co-ordination of hundreds of reactions during cellular activity. Much of the pleasure in studying cells lies in learning to appreciate these.

Cell biology is a large subject and this book is no more than an introduction. Its main aim is to try to sort out underlying patterns from the mass of facts, so that the reader may see the many different facets of cells in perspective. The first chapter sets the scene with a brief review of the whole subject. The remaining chapters are intended to expand selected topics to the point where the reader can tackle the literature himself.

Part One: The Nature of Cells

1: The Cell Theory

Cells and the biosphere

When all the living material on earth is considered as a whole it is called the *biosphere*. With the accumulation of astrophysical evidence it seems virtually certain that there are great numbers of planets in the universe with biospheres, besides our own. The 'living' components of a biosphere are of a highly complex nature and have relatively large amounts of energy built into the chemical bonds holding them together. Consequently there is a tendency for them to decay continuously according to the second law of thermodynamics, which states that in a closed system the components tend to distribute themselves as randomly as possible (or, more precisely, to achieve maximum entropy). To counteract this, energy has to be captured from outside the system (usually radiant energy, for example, from the sun) and this may result in the biosphere remaining in a more-or-less steady state during fairly long periods (millions of years).

Our own biosphere is characterized by the predominance of complex 'organic' molecules based on the element carbon. Solar energy is captured by light-absorbing pigments, such as chlorophyll, and utilized in synthetic processes through the mediation of certain proteins, the enzymes. Enzymes are remarkably specialized macromolecules and there are many thousands of them. They are all made of the same twenty or so amino acids and their specificity is achieved by arrangement of the component amino acids in a specific order. Enzymes themselves decay and therefore for the perpetuation of life a means must exist for their replication. This consists of a synthesizing system programmed by a template molecule which carries information for the assembly of amino acids in the correct order to form a protein. The template molecules are nucleic acids and information is carried in them by the sequence of nucleotides of which they are composed. There are only four kinds of nucleotides in most nucleic acids but, just as the Morse code can carry all the information for all the world's libraries, so they, by forming different combinations, can carry all the information for all the living molecules on this planet. There has to be a mechanism for the replication of nucleic acids also, since they too decay in time. Nucleic acids carry information for their own synthesis since they can split into mirror-image halves and these can each give rise to a mirror-image copy.

For the simplest of living processes to occur it is therefore necessary for dozens of enzymes, nucleic acids and other molecules to be brought together at high concentration in order that they may have a chance to react. Hence, it is a prerequisite of living systems that the components should be spatially confined and prevented from diffusing away from each other. This is achieved by restricting them within lipid envelopes to form cells.

The biosphere is made up entirely of cells and their products. Clearly not all cells are the same and, in fact, the enzymes in the biosphere are distributed in such a way that very few individual cells can carry out all the reactions necessary for their survival. Most cells depend on others and the entire biosphere represents a harmoniously functioning community. Frequently cells are associated in smaller localized communities. Sometimes the association is a very loose one as in a pond where microorganisms, plants and animals exist in balance with each other. Sometimes it is much closer and from such agglomerations may arise colonial organisms and eventually multicellular animals and plants. Secondary phenomena of organization, such as tissue formation and specialization and the development of communications systems, then arise. A yet higher level of organization is represented by the societies of insects and the higher mammals. A common feature of all these systems is the integration and harmonious functioning of the component parts, be these enzymes, cells, organs or organisms. Harmonious functioning depends on interaction of the different components and in living systems special devices are developed for this purpose.

The cell theory

The simplest integrated organization in living systems, capable of independent survival is the cell [117, 458]. This is the reason for considering the cell the basic organizational unit of life and the justification for the study of Cell Biology as a central biological discipline.

The recognition of the cell and the enunciation of the cell theory are among the most recent of human discoveries. Most cells are not visible with the naked eye as they are commonly less than 100 microns long (1 micron (μ) is 10^{-4} cm, i.e. 1/1,000 millimetre). Hence, it was only with the invention of the microscope that observers, such as Borel and Malpighi in the mid-seventeenth century, were able to describe structures which were probably cells. The word 'cell' was itself introduced by Hooke in 1665 to describe the spaces he observed in slices of cork. Many investigators, including Leeuwenhoek, Wolff, Mirbel and Oken, subsequently provided descriptions of cells. Thus the cell theory took shape about the beginning of the nineteenth century, many years before Schleiden and Schwann (in 1838 and 1839 respectively) published their views. These two investigators are, however, usually given the credit for crystallizing the ideas then generally held. Schwann's statement that 'cells are organisms and entire animals and plants are aggregates of these organisms arranged according to definite laws' can hardly be bettered today. It is the concept of the cell as the organizational unit of living systems which will be considered in this book: it will be shown how life has probably arisen as a natural consequence of cosmic evolution and how life processes can be explained in physico-chemical terms.

In adopting this approach we should not entirely lose sight of the fact that, although the biosphere as a whole has not shown any great changes during the past few million years, it has, nevertheless, developed our own human civilization within it during that time. The human species is of particular interest since it can acquire and use information concerning its own physical nature on a vastly greater scale than any other terrestrial form and has recently devised means for carrying terrestrial life to other regions of the universe. The consequences of these properties cannot be clearly foreseen and their interpretation is particularly difficult because it is not easy for us to regard ourselves and all our actions – including reading and writing this book – as manifestations of a special state of matter. There is less difficulty in thinking of cells in these terms.

Common features of cells. The cell can be defined as the smallest organized unit of any living form which is capable of prolonged independent existence and replacement of its own substance in a suitable environment. This definition includes not only animal and plant cells but also the Protista (Protozoa, Protophyta, Bacteria and other microorganisms). It does not necessarily exclude viruses, which may be regarded as either degenerate cells or their primitive precursors.

Cells vary greatly in size and shape. The largest single cells, such as ostrich eggs, single-celled plants like Acetabularia, and nerve cells, may be ten or more centimetres in length. The smallest, microorganisms such as Mycoplasma, are less than 1 μ in diameter. Viruses are smaller still and their dimensions are usually expressed in angstroms. (1 angstrom (Å) is 10^{-4} μ.) They may have almost any conceivable shape.

All cells exhibit some common characteristic features: (i) they utilize extraneous energy to organize atoms and molecules from the external environment and synthesize macromolecules typical of their own structure; (ii) they perpetuate information for their own synthesis through repeated cycles of multiplication; (iii) they control their internal environment in such a way as to create the most suitable conditions for their metabolism; (iv) they regulate their component reactions so that these work in harmony. Moreover, in multicellular organisms cells operate in groups in a harmonious manner.

Major divisions. The *Protista* include bacteria and single-celled animals and plants (Protozoa and Protophyta). *Bacteria* are distinguished by their small size and the absence of a highly-developed internal structure such as is seen in higher organisms [126, 127]. They vary greatly in their metabolic capacities and are commonly divided into groups on this basis. Autotrophs can synthesize all their substance from simple inorganic molecules; heterotrophs depend on products of other living cells. All degrees of heterotrophic requirements are found.

Single-celled animals and plants. These may be divided into the Protozoa (single-celled animals) and Protophyta (single-celled plants), the distinguishing feature between the two being that the Protophyta in general are capable of photosynthesis whereas the Protozoa are not. Many of the Protophyta have a rigid exoskeleton whereas the Protozoa are in general motile. Protophyta also tend to be autotrophic whereas the Protozoa are heterotrophic. In fact, the two groups are not so distinct as this description implies and it is for this reason that they are generally classified together with the Bacteria as the Protista. Many of these single-celled creatures have a highly organized structure which includes nuclei, mitochondria and chloroplasts.

Cells of multicellular organisms. These organisms are the higher animals and plants. The most remarkable feature of cells from multicellular organisms is that individuals may have the same genetic constitution and yet behave very differently. In other respects they have much in common with the Protozoa and Protophyta.

Since, in animals and plants, the functions of the organism are shared among the cells of which it is composed, each individual cell represents an overdevelopment of some function of the totipotent cell. For instance, in many special cells certain general metabolic functions have undergone great development and these are particularly represented in the liver. Other developments of general cellular functions are exemplified by the structural function of connective tissue cells, the protective function of epithelial and endothelial cells and the scavenging function of phagocytic cells.

Some other cells perform highly specialized metabolic functions. Among them are endocrine cells, those which act as carriers, for instance of oxygen, and those which perform special syntheses in the body. Physical functions of cells too may be specialized. Thus, in muscle contractility is highly developed while in nerves conductivity is predominant. In multicellular creatures certain cells also are specialized to carry out reproduction. One of the most interesting and important characteristics of cells from the higher plants and animals is therefore the capacity of the primitive 'totipotent' cell to diverge in many directions in an orderly fashion.

Cells thus constitute a remarkably varied group of organisms. Nevertheless, virtually all of them have certain common properties. For instance, all cells inherit their characteristics from their parents and can synthesize nucleic acids and proteins. Nearly all cells have the capacity to perform anaerobic glycolysis, that is, to break down glucose in the absence of oxygen to yield energy. Nearly all cells use adenosine triphosphate, coenzyme A and the nicotinamide nucleotides in their metabolism. Nearly all cells can concentrate metabolites in their cytoplasm. Furthermore, most cells have common structural features. It is therefore convenient to think in terms of a 'typical cell', usually envisaged as a cross between an animal and a plant cell, with no highly specialized functions, but with the capacity to differentiate in many ways. This convention is a useful one because the hypothetical cell has almost all the properties displayed in any special cell. Almost all cells can be considered as special variants of it or at least as precursors or degenerate forms.

General cellular organization

The great variation in cellular size has already been noted. Small bacterial cells have a volume of about 10^{-13} ml while a large egg may have a volume of 500 ml or more. A typical metazoan cell would come between these extremes and might, for example, have a volume of about $1{\cdot}7 \times 10^{-9}$ ml. Such a cell has a wet weight of about 2×10^{-9} g. Water accounts for about 85×10^{12} atoms of hydrogen and $42{\cdot}5 \times 10^{12}$ atoms of oxygen and the other components of the cell for about $9{\cdot}9 \times 10^{12}$ atoms of carbon, 16×10^{12} atoms of hydrogen, $4{\cdot}6 \times 10^{12}$ atoms of oxygen, $2{\cdot}7 \times 10^{12}$ atoms of nitrogen, $0{\cdot}9 \times 10^{12}$ atoms of sulphur, $1{\cdot}2 \times 10^{12}$ atoms of phosphorus and traces of other elements. Apart from water these atoms form some 350×10^9 molecules of which about $2 \cdot 5 \times 10^9$ are of protein. These molecules make up the structures of the cell; all its functions are manifestations of interactions among them. It is the way in which they are organized in space and time which endows the collection of atoms with the properties of a living cell.

Three separate aspects of organization can be considered in cells. From the *informational* point of view the cell may be regarded as an arrangement for propagation of genetic information, since the ability to reproduce its own kind is one of the most striking features of every organism and is a direct manifestation of the replication of genetic information. Secondly, from the *chemical* point of view, the cell may be thought of as an arrangement for the synthesis of macromolecules from simple molecules in the environment. Certain macromolecules, particularly the nucleic acids and proteins, are quite typical of living protoplasm, and to the chemist the synthesis of these substances is the outstanding property of living matter. Thirdly, from the *thermodynamic* aspect, the cell can be regarded as an open system in which the energy influx is equal to or exceeds the energy efflux so that the energy within the system may remain constant or even increase. Life processes have always challenged the physical chemist because they apparently defy the second law of thermodynamics (which states that the entropy in a closed system must increase to a maximum value). This law implies that the components of a system progressively become degraded to their simplest elements which are eventually dispersed randomly and homogeneously within the system. (Entropy can be thought of as a measure of randomness.) Clearly, living processes achieve exactly the opposite end since small molecules with relatively low bond energies are gathered from the environment, concentrated, and synthesized into larger molecules with higher bond energies and the entropy within the system is consequently decreased. The apparent paradox is resolved by regarding living system as 'open systems' in which energy is constantly entering and leaving so that the system itself remains in a more-or-less steady state.

The three approaches, informational, chemical and thermodynamic may appear to be unrelated interpretations of the same general phenomenon but they are actually closely interconnected and must be considered together in discussing cellular organization. The connections are real ones. Since information within cells is carried mainly by macromolecules, namely nucleic acids and proteins, there is a direct link between the informational and chemical processes. Also, since the synthesis of macromolecules depends on a flow of energy through the cell, the thermodynamic approach is intimately related to the other two.

For these functions to interact at a reasonable rate they must be contained in a restricted environment, and this is, of course, the cell. Even within the cell certain activities are usually restricted to

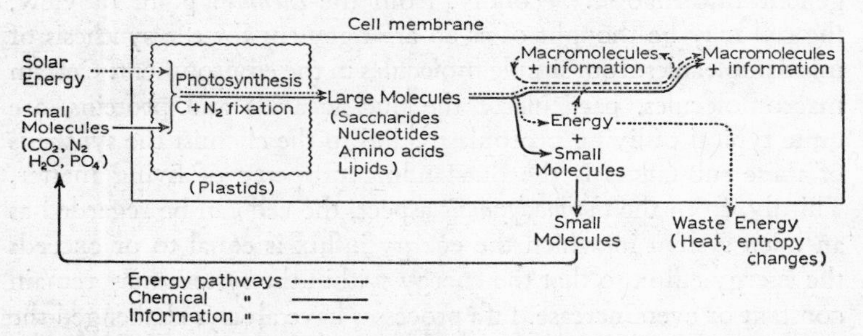

Figure **1.1. Flow diagram to show the relationships among the main energy, chemical and information pathways in cells. Note that the main outcome is an increase of macromolecules and information at the expense of small molecules and solar energy.**

special zones which constitute organelles, such as mitochondria and chloroplasts.

The co-ordinated flow of information, energy and matter within the cell is illustrated diagrammatically in figure 1.1. The essential informational step is the replication of certain macromolecules (nucleic acids and proteins) which carry information in the ordered sequence of their component units. Replication requires that the units (nucleotides and amino acids) should be lined up in correct sequence as dictated by the information already present in the cell and that energy should be available for the chemical linkages. Nucleotides and amino acids are themselves large molecules containing bound energy which can be made available if they and other macromolecules (carbohydrates) are broken down to smaller molecules. This process, indicated on the right of the diagram, represents the most primitive flow pattern of a living system as it occurred in the very earliest heterotrophic protobacteria; today it is represented by the non-photosynthetic organisms.

The complementary part of the process, indicated on the left of the diagram, is the formation of the large molecules which form subunits for macromolecules. These are built up from small molecules, mainly carbon dioxide, water and nitrogen. The energy for their synthesis is derived directly from radiation. In the present world the formation of these large molecules occurs almost entirely by photosynthesis.

The efflux from this system is represented by escape of small molecules formed during the breakdown of large molecules. A certain amount of energy is always lost, of course, since the system is never completely efficient. To complete the picture it has to be added that some random decay is always occurring and this is indicated by the line from 'macromolecules and information' to 'energy plus small molecules'.

This flow diagram is intended to summarize the fundamental properties of living matter. Other manifestations of life can be interpreted as developments of these functions, most of them being involved in the acquisition of raw material or in the removal of waste. Thus the organization of symbiotic cells into communities, the organization of supracellular creatures, the development of motility and of a communications system, may be regarded as evolutionary features which have favoured the survival of certain individuals and their progeny by enabling them to obtain raw materials more effectively. Before proceeding to a detailed consideration of the reactions summarized in figure 1.1, which will be the subject of subsequent chapters, it is proposed to elaborate upon a few of the general features of the system.

Utilization of solar energy. Before oxygen appeared in the atmosphere solar energy was probably utilized for the direct photolysis of water by shortwave ultraviolet light. At the present time short ultraviolet light is filtered out by ozone in the upper atmosphere and energy capture is mainly catalysed by photosensitive pigments such as chlorophyll. In the presence of light these essentially break water into hydrogen ions and hydroxyl ions. The hydroxyl ions recombine to form water and molecular oxygen while the hydrogen ions combine with electrons to form hydrogen atoms: these are used to reduce oxidized or partly oxidized carbon compounds (such as carbon dioxide) to form partly or completely reduced carbon compounds. These are the carbohydrates and lipids which form the sources of energy for all other biochemical reactions. Certain other simple inorganic substances are similarly fixed by reduction with hydrogen. The most important one besides carbon is nitrogen which is reduced to form ammonia and the amino groups of the amino acids.

Large molecules. There are several kinds of these: (i) Carbohydrates and lipids act as energy sources and also provide raw materials for the synthesis of some other substances. (ii) Amino acids provide building units for proteins: they may also act as energy sources. (iii) Nucleotides are the subunits of the nucleic acids: they can be formed by almost all organisms from certain amino acids, carbon dioxide, carbohydrates and phosphates. (iv) Other special large molecules such as the respiratory pigments and the vitamins have also to be included.

When large molecules, such as carbohydrates, are broken down the process of hydrogen fixation may be reversed. Hydrogen atoms may then be removed from a molecule with the release of energy. Energy-yielding reactions of this kind are called dehydrogenations and the hydrogen atoms combine with suitable acceptors, the most common of which is at present oxygen. Energy released in the reaction is rarely dissipated simply as heat. More usually it is incorporated into *high energy phosphate bonds*. The commonest example of a high energy phosphate compound is adenosine triphosphate (ATP). The energy bound in ATP molecules is available for synthetic reactions involved in the formation of nucleic acids and proteins and for other energy-requiring reactions.

Transmission of information. During the synthesis of macromolecules information is fed back into the system. The nucleic acids are the substances involved. Nucleic acids are formed by using the energy available from ATP to link together nucleotides to form polymers. The nucleotides must be linked in a specific sequence and for this purpose the information already present in deoxyribonucleic acid (DNA) is used. Since the original DNA persists through this process, and at the same time a new copy is made, the end result is a duplication of (genetic) information. In the formation of proteins energy from ATP is used to polymerize individual amino acids. Again, the characteristic properties of proteins derive from an orderly sequence of the component amino acids and the information for this is carried by nucleic acids. Consequently information contained in DNA may be used in two ways, first to replicate itself and secondly to produce a specific kind of protein molecule.

Integration and Regulation. Most proteins are enzymes. These are, of course, the organic catalysts which facilitate the many reactions involved in the above scheme. The whole system of reactions, involving hundreds of enzymes in any cell, must be carefully integrated and the control systems are mainly mediated through the enzymes themselves. Three kinds of processes in particular must be carefully regulated: (i) The influx and efflux of materials to and from the cell must be controlled to ensure that materials are available at the correct concentrations for the reactions to take place. (ii) Since intermediary metabolites very rarely accumulate in even very long biochemical pathways, the individual steps in each reaction chain must be co-ordinated. (iii) To maintain the integrity of the whole system, which may be composed of many pathways, these must all be co-ordinated harmoniously. Otherwise, if only a single reaction became slightly out of step with the others it would lead to disruption of the entire organization.

The control processes depend on certain properties of enzyme molecules, two of which will be discussed briefly at this point. The velocity of all enzyme reactions is controlled by two principal factors, (i) the number of enzyme molecules present, and (ii) the concentration of the substrate. Generally speaking there is a direct linear relationship between the amount of enzyme present and the velocity of the reaction catalysed. The relationship with substrate concentration is a little more complex. It is illustrated in figure 1.2 and can be expressed in the following equation.

$$v = k\frac{[E]\,[S]}{1 + K[S]}$$

where v = velocity of reaction, $[S]$ = substrate concentration, $[E]$ = enzyme concentration.

This equation means that when $[S]$ is small (so that the divisor

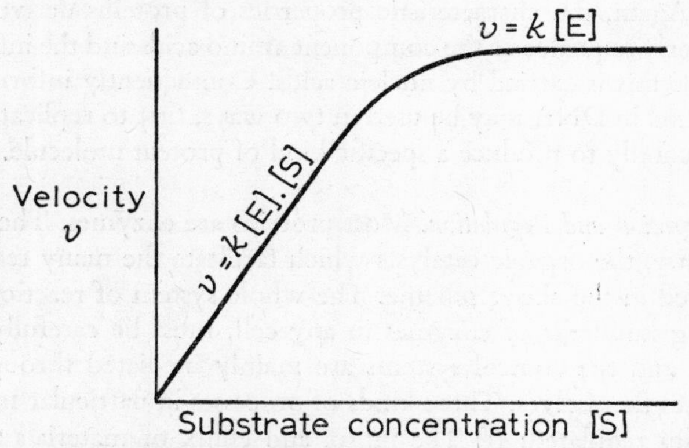

Figure **1.2. The relationship between the velocity of an enzyme reaction and the concentration of the substrate.**

is nearly equal to 1 then, as the substrate concentration is increased, the velocity of the reaction increases almost in direct proportion to it until a maximum value is approached. The velocity then ceases to increase proportionately and eventually (when $[S] = 1 + K[S]$) it ceases to increase at all.

This relationship itself indicates one simple type of control. Consider the following scheme

$$A \xrightarrow{a} B \xrightarrow{b} C$$

Let us consider the reaction $B \rightarrow C$. The substrate for this reaction is B and its concentration is a function of the enzymes a and b. When the velocity of the reactions is equal ($V_a = V_b$) then B is constant, and these are the usual equilibrium conditions. Let us assume that V_a is increased so that B begins to accumulate. Then, if the concentration of B is considerably lower than that required for maximum velocity of the enzyme reaction, the increase of B will lead to an acceleration of V_b. When V_b again equals V_a then a new equilibrium is established. Similarly, if the concentration of B is artificially lowered the velocity of the enzyme reaction decreases and the substrate has an opportunity to accumulate again. Thus a very simple type of control is built into the enzyme molecule.

Most enzymes are also subject to another kind of control which is closely related to the one described. This is *product inhibition*. The term is self-explanatory. If a product accumulates which inhibits an enzyme reaction then it slows down the enzyme reaction and hence the rate of accumulation of the product. Again an equilibrium state is reached, other things being equal. Many enzymes are directly inhibited by their immediate products but it is perhaps even more interesting that some enzymes are inhibited by rather distant products in a chain. This kind of inhibition is referred to as *feedback inhibition* and it forms the basis of the commonest form of control mechanism within the cell.

The two methods of control which have been described are direct controls which enable a system to respond very quickly to changes in the environment. There is yet another kind of control mechanism which is extremely important and will be discussed at length later in this book. In this it is not the catalytic activity of the enzyme molecule which is altered but the amount of enzyme itself. It may increase in response to substrate (*induction*) or diminish in response to product (*repression*).

Cellular structure [83, 124, 145, 153, 391, 432]. The functions we have described can be allocated to particular cellular structures (figures 1.3 and 1.4 and see plates 4, 5). As already mentioned genetic information is stored in deoxyribonucleic acid (DNA) which occurs by itself in lower organisms and combined with protein as nucleoprotein in higher organisms. In these, nucleohistone is organized to form *chromosomes* which are located within the *nucleus*. In bacteria there may be no clearly delimited nucleus and simply a concentration of DNA in one part of the cell (sometimes called the nuclear area). Nevertheless, it is known through genetic studies that this DNA is organized in a specific manner which makes it justifiable to talk of the bacterial chromosome.

Within the nucleus there is another structure, the *nucleolus*, which is often multiple. It is particularly rich in ribonucleic acid and usually disappears during cell division. Its precise function is not yet known.

The protein-synthesizing system is associated with the *ribosomes*,

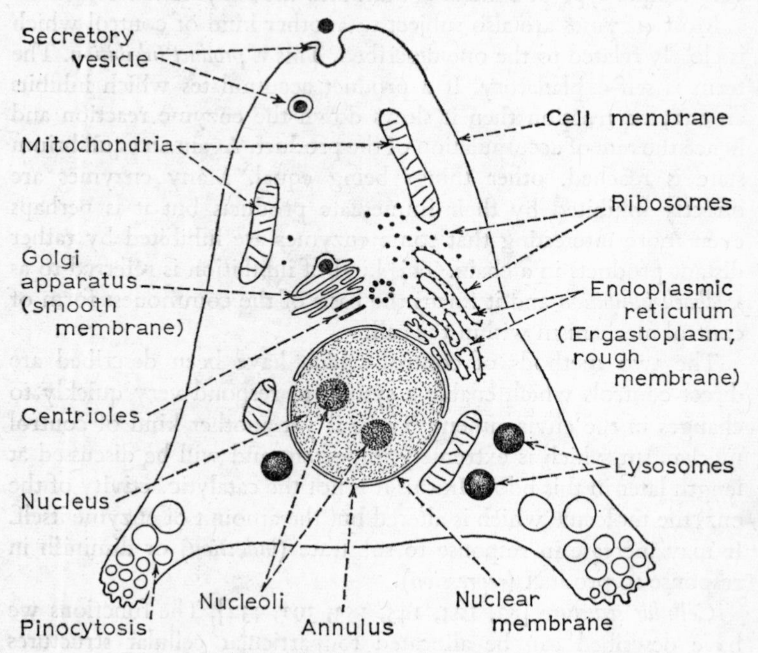

Figure 1.3. Schematic diagram of a 'typical' animal cell.

which in higher organisms are often situated on the *endoplasmic reticulum* (or ergastoplasm). Secretory droplets are present in some cells and these are usually associated with the *Golgi apparatus* [101]. The *dictyosomes* of plant cells are flattened sacs of the Golgi apparatus.

The enzymes and pigments involved in photosynthesis are located in *chloroplasts*. The energy-yielding systems involved in the synthesis of ATP and the oxidation of reduced carbon compounds are in the *mitochondria*.

Many cells contain digestive vacuoles. In animal cells these are probably represented by the *lysosomes*. These bodies contain proteases, nucleases and other catabolic enzymes [27]. In the cytoplasm of all animal cells capable of division are found the *centrioles*, whose precise function has not yet been determined. All these structures include enzymes and many enzymes are also found free in the cell

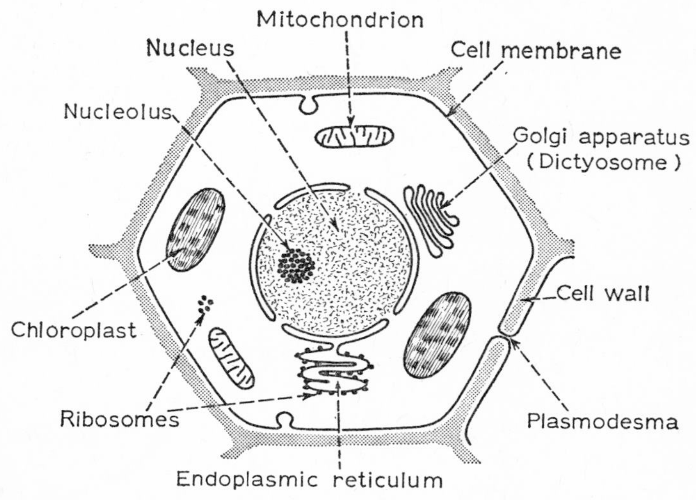

Figure 1.4. Schematic diagram of a 'typical' plant cell.

sap. Other specialized inclusion bodies are found in certain cells and include storage material such as glycogen and lipid. The whole cell is enclosed within a *cell membrane* and individual organelles are also separated by membranes such as the *nuclear membrane* and the *mitochondrial membrane*. The membranes may be modified to give rise to villi, flagella, cilia and other special structures.

Outside the cell membrane lies the cell wall which is very prominent in most plant cells but is often hardly detectable in animal cells. Adjacent plant cells are often connected by cytoplasmic bridges, *plasmodesmata*, which penetrate the cell wall.

Quite recently another structural element in the form of microtubules has been recognized in most cells [119, 282, 417, 464, 470, 514]. These are associated with the mitotic apparatus, the subcortical zone in plant cells and many other structures. It is thought that they may be contractile.

Part Two: The Molecular Basis of Cellular Structure

2. Macromolecules

Macromolecules (giant molecules) are the characteristic structural components of living matter [32, 150]. These large molecules are polymers made up of unit molecules linked to each other in a special manner. The most important are the nucleic acids, the proteins and some carbohydrates such as cellulose and starch. Lipid and lipoprotein membranes are in many respects similar structures but, since they have special features and are held together principally by adsorption forces rather than by valency bonds, they will be considered separately in connection with membranes of which they form the major component.

In nucleic acids and proteins the units which make up the polymers are not connected in a random fashion but are ordered in a particular sequence, upon which the most important properties of the molecules depend. The proteins, consisting of some twenty amino acids, include all the enzymes which catalyse the vast majority of all biological reactions. The nucleic acids, mainly composed of only four nucleotides, carry the code which dictates the order in which amino acids appear in proteins. The properties of these two kinds of molecules depend on the precise sequence of the monomers (sometimes hundreds) of which they are composed. They are therefore unique among chemical substances. They are also somewhat unusual in their large size. The smallest macromolecules, some of the proteins, are 20 to 100 Å in diameter while the largest ones, molecules of deoxyribonucleic acid, may be hundreds of microns long. Some characteristics of these molecules will be described in this chapter.

Nucleic acids

The nucleic acids [81, 110] comprise deoxyribonucleic acid (DNA) and ribonucleic acid (RNA) of which there are several distinguishable kinds. They are concerned almost entirely with conveying and replicating genetic information. Both DNA and RNA are made up of nucleotides which themselves are composed of a base, a sugar and phosphate (figure 2.1). The bases are either purines

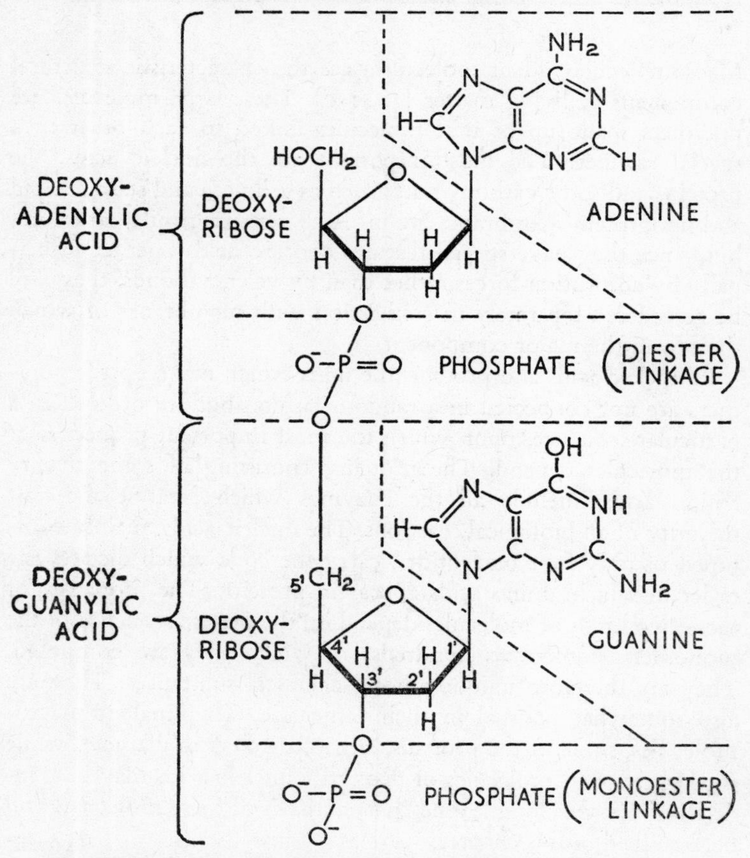

Figure 2.1. Structure of a dinucleotide, composed of deoxyadenylic acid and deoxyguanylic acid linked by a 3′ : 5′ diester linkage, as in DNA.

or pyrimidines. The most abundant purines, which occur in both kinds of nucleic acids, are adenine and guanine. Of the pyrimidines cytosine is found in both but thymine occurs only in DNA and uracil almost exclusively in RNA. A few rarer bases occur in RNA, almost solely in soluble RNA which will be discussed later. The sugars are pentoses, deoxyribose in DNA and ribose in RNA.

In the formation of a polynucleotide, pairs of nucleotides are joined by phosphate diester linkages between the 3-position of the pentose of one nucleotide and the 5-position of the pentose of its neighbour. A series of nucleotides may be linked together in this way to form a backbone of alternating sugar and phosphate residues from which the bases radiate. In order to facilitate description of polynucleotides two shorthand notations are used. In both of these letters represent individual bases: A for adenine, C for cytosine, T for thymine and so on. The letter p is used to indicate a phosphate group. In the first notation the letters are joined by a zig-zag line as shown in figure 2.2. Horizontal lines symbolize sugars and

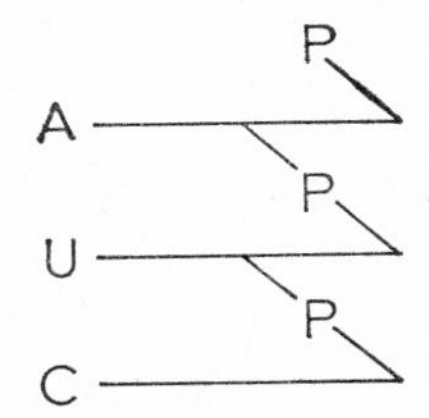

Figure **2.2 Diagrammatic representation of a polynucleotide.**

diagonal lines indicate the way in which they are linked. Hence when the diagonal line meets the end of the horizontal line it indicates a 5-linkage whereas when it meets the middle of the horizontal line it indicates a 3-linkage. In the second type of notation the lines are eliminated altogether and the formula is simply written in the form pApUpC and so on. In this notation when p comes before a letter it indicates that the phosphate is attached to the nucleoside by a 5-linkage.

Deoxyribonucleic acid. Analysis of DNA revealed two very striking features. In the first place chemical analysis demonstrated that certain bases are always found in the same amount as certain others. Chargaff and his colleagues [80] first demonstrated that the amounts of adenine and thymine in a given specimen of DNA are always approximately equal and that the amounts of cytosine and guanine are also equivalent. The other striking property of DNA demonstrated by Wilkins and his colleagues [510] is that it gives a pattern in X-ray diffraction studies which indicates a regular helical structure with certain well-marked features.

On the basis of this information Watson and Crick [492] pro-

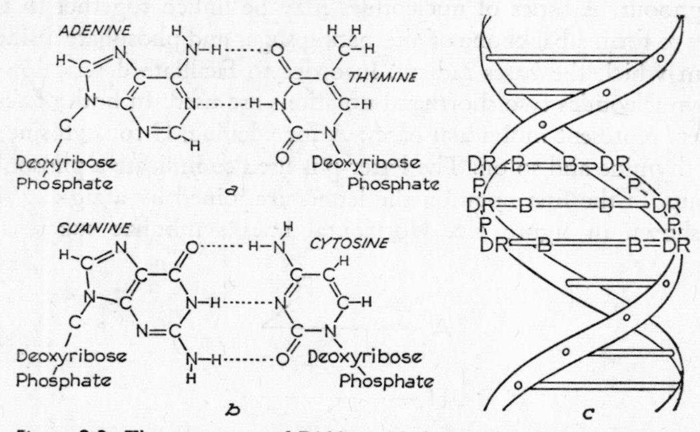

Figure **2.3. The structure of DNA. a. Hydrogen bonds linking adenine and thymine. b. Hydrogen bonds linking guanine and cytosine. These relationships are responsible for specific pairing between pairs of bases. c. The way in which nucleotides are joined to form a double helix. In each strand the nucleotides are linked by 3′ : 5′ phosphate diester linkages; the strands are held together by hydrogen bonding between complementary bases.**

posed that the DNA molecule consisted of a double strand of nucleotides in which the bases of one strand were always paired with the bases of the other strand by means of hydrogen bonds. Adenine and thymine could fit together in this way and cytosine and guanine could link similarly to give the correct spacing (figure 2.3*a*, *b*). These linkages formed the basis of the right-handed helix which forms the familiar spiral staircase model (figure 2.3*c*). It may be noted that this double spiral has two alternating grooves on its surface, one broad and the other narrow. The resulting molecule is about 20 Å n diameter and may be very long. Indeed if certain viruses and chromosomes consist of a single molecule of DNA (as seems very likely) then they may extend to many microns. Each complete turn of the helix increases by 34 Å and there are approximately 10 base pairs in each turn.

One extraordinarily interesting feature of the DNA molecule is

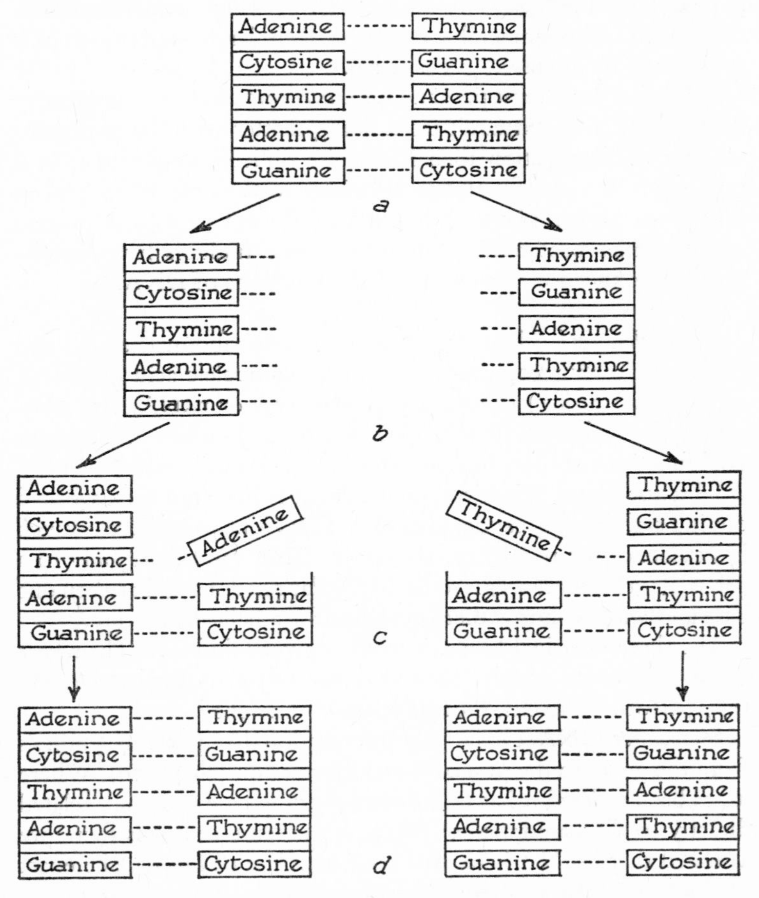

Figure 2.4. The way in which DNA replicates by pairing complementary bases with the bases in each individual strand.

that its structure immediately suggests a mechanism for its own duplication (figure 2.4). If the two strands are separated (and in theory this is not difficult since they are held together only by hydrogen bonds) then complementary halves are obtained. If each one then acts as the template for the synthesis of a complementary strand by pairing of individual nucleotides, the end result is two complete double helical molecules each of which is identical with the parent. Experimental evidence which lends support to this model will be discussed in chapter 6.

One other important point about DNA remains to be mentioned. If care is taken to extract it from living matter with the minimum amount of degradation then it is quite difficult to obtain it in a pure form. It is nearly always associated with some RNA and in cells from higher plants and animals is bound to a basic protein usually either protamine or histone. These findings have significance in connection with some of the functions of DNA.

Ribonucleic acid. There are several kinds of ribonucleic acid and three are clearly recognized: messenger ribonucleic acid (mRNA), soluble (or transfer) ribonucleic acid (sRNA or tRNA) and ribosomal ribonucleic acid (rRNA). There may be others. The individual functions of these nucleic acids have been well established and are discussed later. However, their structures have not been worked out in nearly such great detail as that of DNA because there is much less evidence for regularity. Except in tRNA (and there only to a partial degree) there seems to be no regular relationship between one base and another nor is there evidence for a helical structure.

The three principal kinds of RNA are very different. Their base ratios may differ widely, they vary greatly in molecular size and they perform different, though related, functions.

Messenger RNA carries the information for protein synthesis from the genes to the sites of protein formation (chapter 6). Where it has been isolated it has been found that the base ratios are related directly to the base ratios of DNA in the same cell; it is therefore assumed that mRNA is formed on a template of DNA. The exact manner in which it is copied is not established with certainty but it seems most likely that mRNA is formed simply by base pairing with one strand of DNA, adenine pairing with uracil, cytosine with guanine, and so on.

Ribosomal RNA, as its name suggests, is found in the ribosomes, small particles with a molecular weight of about four million which occur in the cytoplasm of almost all cells. They are composed of RNA and protein, about 60 per cent being RNA. The base ratios are very similar in ribosomes from many different microorganisms,

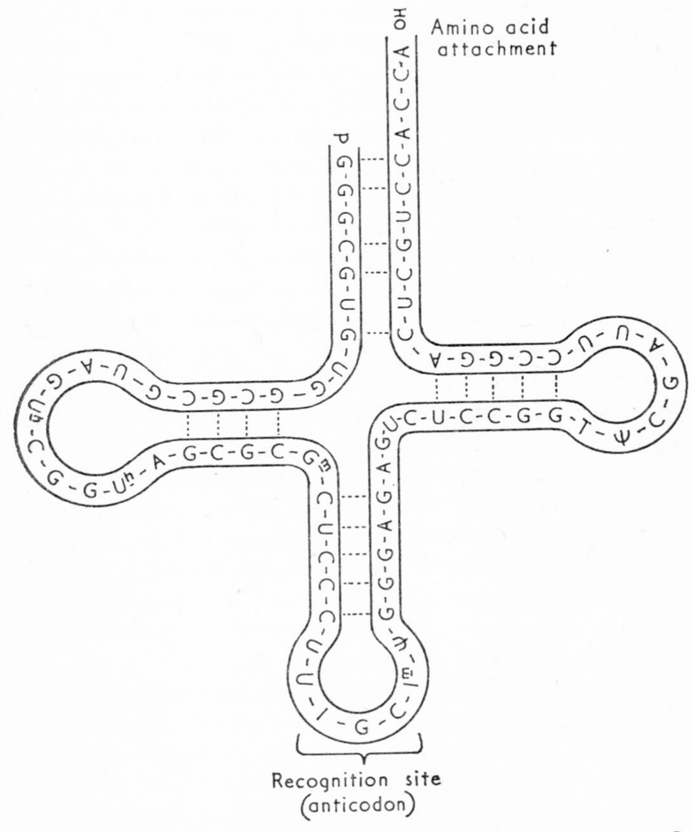

Figure 2.5. The structure of alanyl-transfer RNA. A – adenosine, G – guanosine, C – cytidine, U – uridine, U^h – dihydrouridine, I – inosine, ψ – pseudouridine, ^{m} – methyl-.

suggesting a general structural similarity. No regular relationship among the bases themselves has been recognized and no regular structure has yet been detected.

The most precisely characterized RNA is tRNA (transfer RNA) which has the special function of acting as an adaptor for attaching amino acids to the mRNA template in the course of protein synthesis [214]. It is so constructed that it can combine with an amino acid at one part and with the specific region of an RNA template at another part. tRNA molecules are relatively small for nucleic acids, having a molecular weight of some twenty to twenty-five thousand which corresponds to between sixty-five and seventy nucleotides. All molecules of tRNA have the same terminal sequence (cytidylyl-cytidylyl-adenylic acid) and contain certain unusual bases which are not found commonly in other nucleic acids. The commonest are pseudouridylic acid and 5-methylcytosine. In certain conditions a considerable part of the molecule may consist of double helices [444] formed by a single strand doubling back upon itself (figure 2.5). In one of the unpaired regions three bases can be identified which are known to be complementary to the three bases constituting the genetic code for an amino acid. These are thought to constitute the recognition site for mRNA, that is, the site at which it becomes attached to a template. There are individual tRNA molecules for each amino acid (sometimes more than one) and therefore there are quite a large number of them, probably just over forty in most cells. The sequences of the nucleotides in some tRNAs have now been determined (figure 2.5) [221, 222]. Common structural features have been revealed in several.

Proteins

Proteins [340, 341] in living cells perform two main kinds of functions, catalytic and mechanical. In the form of enzymes they catalyse most of the chemical reactions in living systems. Mechanical functions include structural functions of the kind performed by collagen and elastic tissues, and also contractile functions such as are performed by muscle protein.

Proteins are polypeptides (polymers of amino acids). There are approximately twenty amino acids most of which have the general formula RNH_2COOH, where R is a specific side group. These are linked together by the peptide linkage, —CONH—, which is formed by the loss of a molecule of water between two amino acids. The major exception to this general structure of amino acids is proline (and also its derivative hydroxyproline). Proline is an imino acid and it is of special importance in connection with the secondary structure of protein molecules as will be described later.

Primary structure. It is convenient to think of protein structure at four levels of organization: primary, secondary, tertiary and quaternary. The primary structure of a protein is simply the order

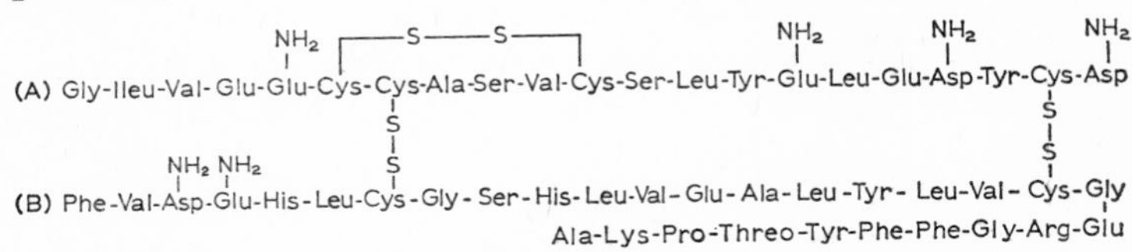

Figure 2.6. The primary structure of insulin.

in which the amino acids occur within it and primary structures of several proteins have now been worked out in detail. The simplest of these is the hormone insulin (figure 2.6) whose structure was worked out by Sanger and his colleagues [418].

Secondary structure. A polypeptide chain can be considered to have a skeleton consisting of a sequence of nitrogen and carbon atoms from which hydrogen atoms and side chains radiate. A unit section of this chain is shown in figure 2.7 and it can be seen to consist of alternating amide groups and α-carbon atoms of amino acids. The atoms of the amide groups tend to occur in the same plane because of the bond angles and they can be distorted out of this plane with difficulty. On the other hand the amide groups can rotate fairly freely about the α-carbon atoms. Because of this repeated pattern in the skeleton of the molecule there is a tendency for it to assume a stable three-dimensional structure and this is the *secondary structure.*

The simplest example is the so-called 'pleated sheet' arrangement such as is found in silk fibroin [308]. Successive planar amide groups are arranged at right angles to each other so that the molecule, when viewed from the side, appears to form a zig-zag or pleated structure as shown in figure 2.8. Proteins exhibiting this

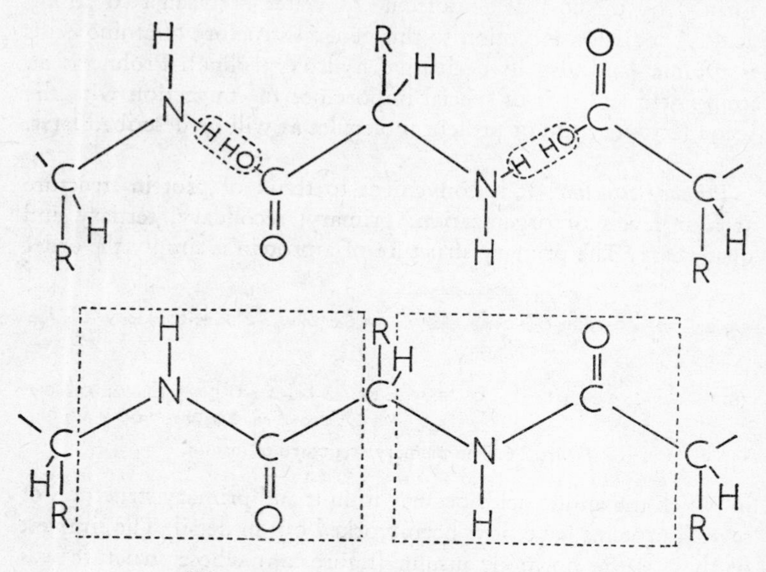

Figure 2.7. Peptide bonds between three amino acids giving rise to two planar amide groups. These are rigid and resist distortion.

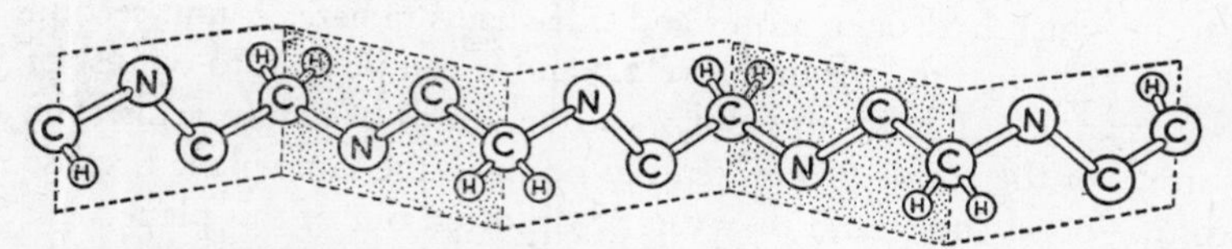

Figure 2.8. The pleated sheet arrangement, showing how it is made of alternating planes.

structure generally have a rather abnormal composition. For instance, fibroin contains 44 per cent of glycine, while most of the remaining amino acids are alanine and serine. These are all amino acids with very small side chains which can permit the pleated arrangement to form and remain stable.

In most other amino acids the side chains are too large and would cause serious distortion of any pleated arrangement but another kind of stable three dimensional structure can be formed to accommodate them. This is the so-called α-helix [92, 368, 369, 370, 371] which is illustrated in figure 2.9. If this figure is carefully studied it may be seen that the atoms of each amide group still lie in a common plane and that rotation occurs about the α-carbon group to form a right-handed spiral structure. The spiral is stabilized by hydrogen bonds joining the NH group of each amino acid to a CO group of an amino acid in the next spiral. Each turn of this spiral increases by 5·4 Å and contains 3·7 amino acids on the average. Since the side-groups protrude from the outside of the molecule and are separated by some 5 Å from the nearest neighbouring side-groups, it is possible to accommodate almost all the possible amino acids in this configuration. However, one particular component of proteins cannot fit into either the pleated sheet arrangement or the α-helix. This is proline (or its derivative hydroxyproline) which is an imino acid in which the amino group doubles back to react with the δ-carbon atom to form a closed ring. Such a configuration cannot fit in with the regular structures which have been described. This property of proline is particularly important in relation to the tertiary structure of certain molecules which will now be discussed.

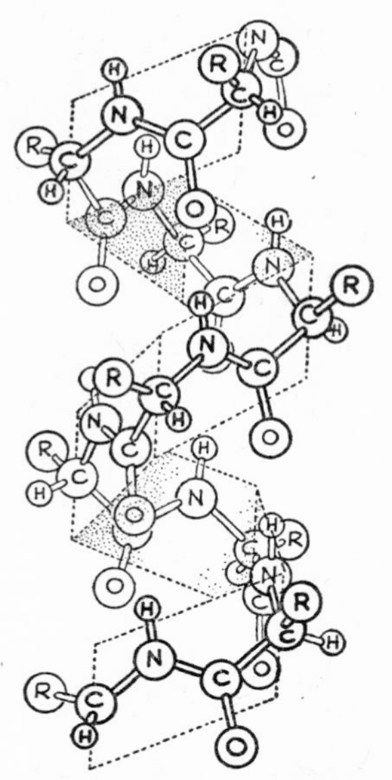

Figure **2.9. The α-helix is made of a series of planes arranged to form a right-handed spiral.**

Tertiary structure. Superimposed on the secondary structure of proteins there may be another order of organization. The simplest example is a super coiling of the α-helix such as occurs in keratin to give a coiled coil structure, like an electric lamp filament. Of special interest are the detailed studies which have been made of the tertiary structures of the myoglobin and haemoglobin molecules by X-ray diffraction [265, 266, 267, 383]. The structure on the left hand of plate 1 represents the myoglobin molecule according to Kendrew and his colleagues. In this molecule most of the straight sections consist of α-helices. It contains several proline molecules and these are found to occur at angles. This structure, which at first sight might appear haphazard, is in fact very stable owing to hydrogen bonds and other cross-linkages.

The haemoglobin molecule, shown also in plates 1 and 2, has a rather similar structure to myoglobin. Indeed many of the sections of both molecules may be identical. Full consideration of the structure of haemoglobin leads to the realization that there is yet a further degree of organization.

Quaternary structure. Each complete haemoglobin molecule is composed of four unit macromolecules, one pair of each of the

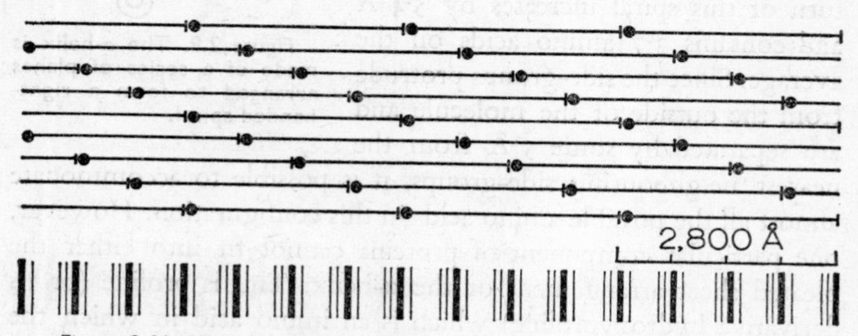

Figure **2.10. The arrangement of tropocollagen units to give a collagen fibre. Because adjacent tropocollagen molecules overlap a quarter of the length of a single molecule the regular repeating pattern at the bottom is obtained and this can be seen with the electron microscope.**

structures shown in the centre and on the right of plate 1. As shown in plate 2, two pairs of each submolecule can be fitted together to form a tetrahedral arrangement. They fit together so closely that there are very large areas of contact between them.

Another important example of quaternary structure is the molecule of collagen [191, 421, 477]. Collagen is made up of a very large number of sub-units of tropocollagen which are linked together in the manner shown in figure 2.10 to form the fibres of collagen. These show a characteristic structure with the electron microscope because of the regular repetition of corresponding areas (plate 3A).

There are many other examples of quaternary protein structures in which the details of association of the component molecules have not yet been worked out. In particular the regular structure of many viruses is due to the repetition of protein sub-units in a quaternary structural arrangement.

Compound molecules. Analogous to the quaternary structure of certain proteins, and at the same level of organization, is the structure of many of the compound molecules which are formed by association of protein molecules with molecules of other substances. These compound molecules include nucleoproteins, lipoproteins and glycoproteins. Nucleoproteins are formed by the association of nucleic acids with proteins [137, 497, 500]. The reaction probably occurs between phosphate groups on the outside of the nucleic acid molecule and basic groups on the protein molecule. Examples of nucleoprotein structures are the ribosomes, certain viruses and the DNA-histone complexes found in higher plants and animals. These substances belong to the class of coacervates [49]. Lipoproteins form a group by themselves and are of particular interest since they form most of the membranes of cells.

Because of the enormous numbers of possible combinations of amino acids it can be appreciated that the potential number of proteins in nature is astronomical. A rough equivalent would be the number of sentences which could be written employing the twenty-six letters of our alphabet. Very many different proteins are, in fact, found in nature but they can be classified into a relatively small number of groups [14]. For instance, most proteins are enzymes and the number of kinds of enzymes (e.g. esterases or lactic dehydrogenases) is relatively small. On the other hand, similar proteins from different species almost invariably exhibit unique features owing to the fact that, although the active site may be similar in all the sequence of amino acids in many parts of the molecule may differ from one species to another. Even within a species the same reaction may be catalysed by several enzymes. Enzymes from different species which carry out the same reaction are referred to as heteroenzymes while enzymes from the same species which carry out the same reaction are referred to as isozymes or isoenzymes [307]. Some isozymes may simply represent heteroenzymes which have persisted in different species during evolution. Others, however, seem to be specialized and may be functionally different in some particular respect; for instance, in their response to feedback control.

Single protein molecules sometimes have more than one specialized site, and, indeed, some can catalyse two different reactions. Even when a protein catalyses only one reaction it may nevertheless have several specific structural sites; for example, the catalytic site, a site which ensures substrate specificity, and special configurations to permit inhibition by feedback from distance products. Some molecules probably have special structures to facilitate their attachment to cell membranes. There may also be large areas of protein molecules whose sole purpose is to stabilize them. There is good evidence that any or all of these structures may vary without causing enzymatic activity to be lost and this no doubt accounts in no small part for the enormous variety of antigens found throughout nature.

Polysaccharides

An account of macromolecules in living cells would not be complete without mentioning polysaccharides. These are polymers made up of monosaccharides such as glucose, hexuronic acids and amino sugars. They perform three main functions in cells. In the form of glycogen and starch they are the principal stores of chemical energy in the organism and provide the primary fuel for most biochemical reactions. As cellulose, chondroitin and similar substances, they have important mechanical functions and constitute the main extracellular structural materials of most higher organisms. As hyaluronic acid and, in combination with proteins, as mucoprotein they also have a general protective function at the surfaces of cells. They may behave as cation exchangers and hence act as a primary ionic barrier. Another important biological property of polysaccharides (particularly mucopolysaccharides) is that, like proteins. they can be antigenic.

3. Biological Membranes

Lipids and Lipoproteins

The two most important groups of lipids in biological materials are steroids and fatty acid esters, mainly glycerides and phospholipids.

Two groups of steroids are of particular interest. *Cholesterol* occurs predominantly in membranous structures in cells. The *steroid hormones* participate in the control of metabolic processes and will be considered in chapter 11. Of the fatty acid esters, the glycerides are mainly important as storage substances while the phospholipids occur principally in membranes [113]. Cholesterol and certain of the phospholipids are of particular interest in relation to membranes; the formulae of some members of this group are shown in figure 3.1.

Phospholipid molecules have a long hydrophobic chain, with a small polar and therefore hydrophilic group at one end. When substances of this kind are allowed to spread out on a liquid surface the molecules tend to pack together, with the polar groups towards the liquid and the hydrophobic groups sticking outwards forming an orientated monolayer [105, 108, 118, 251]. With monolayers of phosphatidyl choline (lecithin) it is possible to measure the area occupied by each molecule and this is about twice the cross-sectional area of the hydrocarbon chain. When cholesterol is added to a phosphatidyl choline monolayer it causes it to collapse, presumably owing to the formation of a more closely packed system of the arrangement shown in figure 3.2. This kind of structure probably forms the basis of lipoprotein membranes in cells.

Monolayers of phospholipid on water may, on becoming

hydrated, form thread-like tubular structures described as myelinic or myelin forms. These can be studied by both X-ray diffraction and electron microscopy [231, 401] and have been found to consist of bimolecular leaflets 60–120 Å in thickness. Their resemblance to biological membranes is striking, and most biological membranes are probably formed of bimolecular lipid leaflets of this nature. Biological membranes also contain protein and Danielli [106, 107]

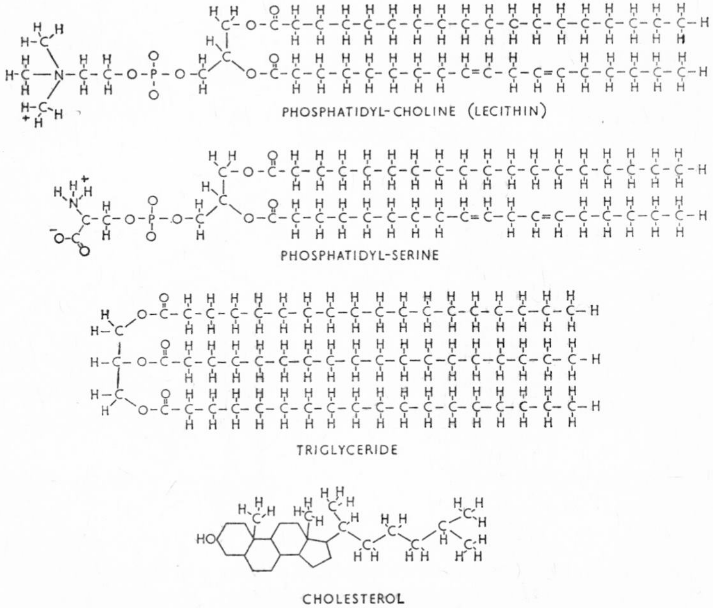

PHOSPHATIDYL-CHOLINE (LECITHIN)

PHOSPHATIDYL-SERINE

TRIGLYCERIDE

CHOLESTEROL

Figure **3.1. The chemical structures of some of the lipid molecules which occur in membranes. These are shown approximately to scale.**

suggested that the basic membrane is a bimolecular lipid leaflet with the hydrophobic groups opposed and the ionic groups turned outwards where they can react with charged groups on proteins. The unit membrane is thus visualized as a kind of double sandwich with protein on the outsides and two opposed lipid layers inside [408, 409, 410].

The most detailed studies of membrane structure relating to these properties of lipoproteins have been made with the myelin sheath of nerve [124, 140, 142, 144, 145, 163, 420]. This sheath is formed by the wrapping of Schwann cells round and round nerve axons in the manner of a swiss roll [177]. Almost all the cytoplasm is squeezed out of the membranes so that the myelin sheath eventually consists of concentric layers of Schwann cell membrane and is there-

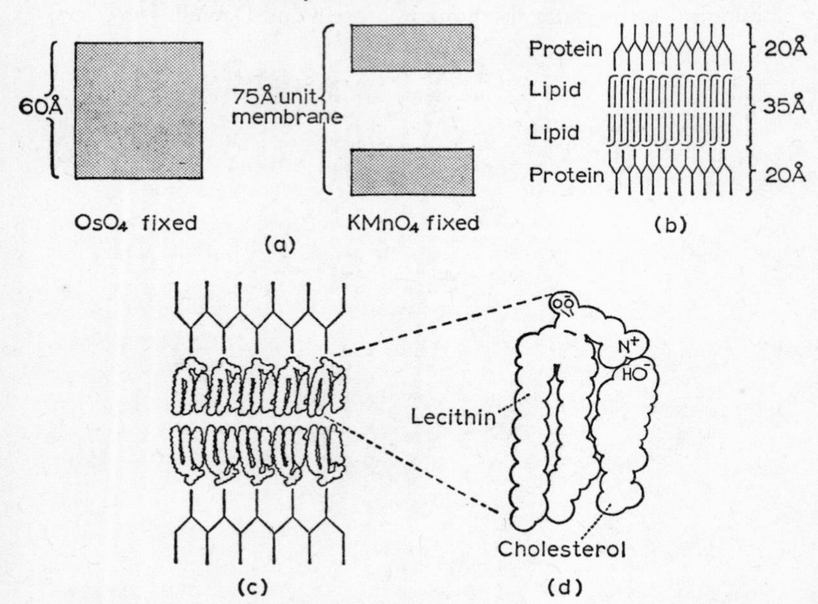

Figure 3.2. The hypothetical structure of a lipoprotein membrane built up on a double layer of lecithin and cholesterol. a. The appearance of a unit membrane in the electron microscope. b. The relationships of the major components of a unit membrane. c. The manner in which the lipid units pack together. d. The manner in which lecithin and cholesterol become associated. These are the actual shapes of the molecules.

fore a particularly suitable object for study. In electron microscopical section (plate 3B) it resembles closely the description of a myelinic form and consists of heavily osmiophilic bands about 30 Å thick separated by lighter areas about 110 Å wide in the middle of which lie bands of intermediate density. In such sections the distances are rather smaller than those indicated by X-ray diffraction, owing to dehydration in the course of fixation. The pattern is compatible with a structure of double membranes each consisting of a bimolecular lipoprotein leaflet, the membranes being separated by the band of intermediate density which is the site of the so-called 'difference factor'. Chemical analysis of this membrane has shown it to contain phospholipid, cholesterol and cerebroside in the ratios 2 : 2 : 1.

Membranes

In the simplest cells, such as bacteria, there may be only an outer limiting membrane but in more highly developed cells there are many membranous structures [132, 234, 412, 433]. The principal ones are the cell membrane itself, the membrane of the nuclear envelope, the membranes of the endoplasmic reticulum and the membranes of organelles such as mitochondria, chloroplasts and lysosomes.

Two general functions are common to nearly all membranes. One is to restrict the diffusion of molecules and ions. The other is to provide a structure on which proteins may be assembled. For instance, in the mitochondrion respiratory enzymes are arranged on the membrane in a specific order. The lipoprotein membranes of the cell therefore merit the term 'cytoskeleton' which was originally proposed to describe a hypothetical structure of similar nature.

The cell membrane

The cell membrane is also known as the plasma membrane or plasmalemma. Until fairly recently arguments were advanced that there was no definite membrane at the cell boundary but that the phenomena attributed to a membrane could be explained by the behaviour of an interphase between the environment and the cytoplasm. However, the evidence for a membrane is now conclusive.

The plasma membrane is itself covered with a coat, the cell wall or cell coat. In many animal cells this is barely detectable and its existence is sometimes a matter of dispute but in plant and many protistan cells it is very prominent. The properties of the cell membrane may be profoundly modified by this coat and its chemical structure is usually distinctive of particular cells or tissues.

Properties of cell membranes

Permeability [112]. It was shown many years ago that the ease with which substances pass through the cell membrane depends on three factors.

(i) In general, the greater the solubility of a substance in lipid solvents the more readily it passes through the cell membrane. For a series of substances the rate of passage can be directly correlated with the partition coefficient between lipid solvents and water [88]. This is true irrespective of the size of the molecules, provided they are fairly large.

(ii) Certain very small ions can pass readily through the cell membrane. On the whole they must have ionic radii of less than 7 Å.

(iii) Some compounds cannot be covered by any general rule. Into this category fall substances such as carbohydrates and proteins on the one hand and sodium ions on the other. For all these there is evidence for special mechanisms to facilitate their passage through the cell membrane.

Surface tension. Because of the behaviour of cell membranes towards lipid-soluble substances it is naturally assumed that the membrane is of lipid nature. In this event, however, the membrane would be expected to exhibit very high surface tension. In fact the surface tension is very low and this is strong evidence for the adsorption of protein to the outside of the lipid layer [200].

Electrical properties. The cell membrane exhibits a rather high electrical resistance and this again seems to provide evidence that it consists largely of lipid material [86, 258].

Structure of the cell membrane (figure 3.3). Although precise details are still in dispute it is generally agreed that the basic structure of a typical cell membrane is almost certainly a double layer of phospholipid with the non-polar groups apposed and the polar groups turned outwards [107, 436]. In more highly developed cells the phospholipid layers are almost certainly compound layers of cholesterol and phospholipid as already described [461]. (Bacteria do not contain cholesterol.) To explain the free entry of very small particles (<7 Å) it is proposed that this lipid layer is penetrated at intervals by pores with a diameter in the region of 7–10 Å. These pores are probably lined by positively charged groups since cations are specifically excluded. Attached to the lipid layers are probably protein layers both on the outside and the inside. The outer layer may include mucoproteins containing sialic acid [91, 212, 363, 499]. Other proteins associated with the membrane include enzymes [234, 353] some of which may be involved in transport of certain substances, such as sugar and sodium, across the membrane. In the

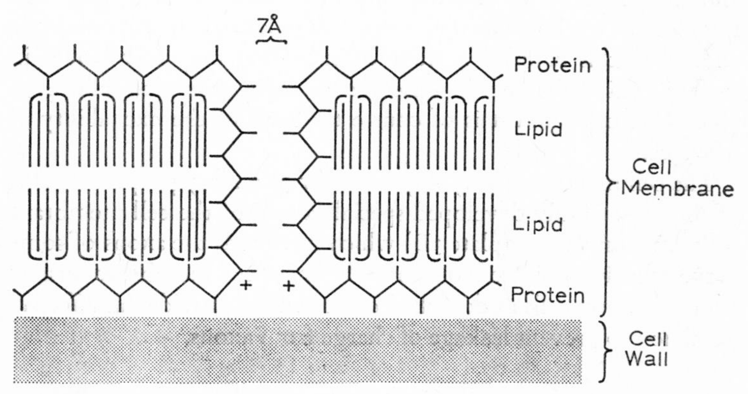

Figure **3.3. A diagram of the structure postulated for the cell surface.**

case of sodium a special mechanism, the sodium pump, is involved and one of the most plausible theories suggests that this is located in proximity to the 7 Å pores.

Special functions of the cell membrane. Details of the sodium pump will be discussed in a later chapter but at this point some remarks may be made about the entry of larger molecules. The actual mechanism whereby glucose, for example, gains entry to the cell is not understood. There is conflicting evidence as to whether energy is required but it seems quite likely that, in the presence of a suitable gradient glucose can enter or leave freely without the intervention of any energy-utilizing reaction. The mechanism suggested by

Mitchell [322, 323] for this kind of transfer is the so-called translocation process which proposes that the molecule to be transported fits into a siteof special configuration on the outside of the membrane and is then carried through the membrane and released on the other side. It is postulated that the carrier site is highly stereospecific for each molecule and that only when the site is filled can the carrier perform its function.

Another property of the cell membrane, conductivity, is important. By virtue of the ionic pumping mechanisms different concentrations of ions are established on either side of the cell membrane and this results in an electrical potential across it [423, 469]. Cell membranes are however quite easily depolarized by minor stimuli and when this happens a wave of excitation may travel over its surface. This behaviour is observed in many kinds of cells, even plant root cells, and it is of course the basis of the excitability of nerve cells.

The cell membrane may acquire other secondary properties. One of the most highly developed specializations of the cell membrane is the myelin sheath (plate 3B) which surrounds the axons of some nerve cells. It consists of concentric layers of cell membrane formed from Schwann cells as described earlier and it provides an effective insulation to prevent leakage of charge from axons.

The external cell coat

The external cell coat [354, 355] has multiple functions. In plants and bacteria it provides osmotic protection. For instance, bacteria which normally live in very dilute solutions can be treated with enzymes which remove the cell coat. The denuded cells are called protoplasts. These burst if placed in dilute solutions and can be kept intact only in solutions of quite high osmotic pressure.

The cell coat also provides the principal structural material of many organisms. Thus the main cell wall materials in plants are substances such as cellulose and lignin. In most animals too the cell coat has a structural function as a principal component of the connective tissues. For example, chondroitin sulphuric acid, hyaluronic acid, collagen and similar substances perform essentially the same functions in some animals as cellulose and lignin do in plants. Similar materials probably coat most cells and they may also perform other functions, such as molecular filtration and ion exchange.

Characteristically cell walls have a high carbohydrate content and contain acetylated sugars and uronic acids. Some of the typical cell wall materials are entirely polysaccharide and others such as lignin and collagen are usually formed secondarily and in association with the mucopolysaccharides.

In plant cells pectin is usually the first material to be formed at the surface. It is polygalacturonic acid. Subsequently cellulose is generally formed. It is polyglucose, in which the glucose molecules are joined to each other in such a way that the molecule packs into a tight spiral which forms a long rigid structure. These long molecules give cellulose its distinctive property of rigidity. In lower plants another polysaccharide, chitin, commonly occurs. It is poly-N-acetylglucosamine. In 'woody' plant tissues lignin is eventually formed. It is not a carbohydrate but is a polymer made up of aromatic compounds.

In animal cells a parallel situation is found. Commonly the earliest formed intercellular material is hyaluronic acid which is a polymer containing acetylglucosamine and glucuronic acid. Subsequently other polysaccharides, such as chondroitin sulphuric acid may appear. Chondroitin sulphuric acid contains acetylgalactosamine and glucuronic acid and has sulphate groups attached. Eventually proteins, such as collagen and elastin, are laid down. A common component of many kinds of cell walls has recently been shown to be sialic acid (or neuraminic acid), also a carbohydrate. Its significance is not understood but it is suspected to play a particularly important part in stabilizing the cell membrane. Chitin, already mentioned in connection with plants, is also found in many animals, for instance insects. As has already been mentioned, polysaccharides may act as ion exchange materials and reduce the diffusion of certain ions. Because of their lattice structure they may also act as molecular filters and exclude large ions. For instance serum proteins are filtered out from cartilage cells and from many other connective tissue cells in this manner.

The main function of the cell membrane and cell wall is, then, to maintain the internal environment of the cell. This is achieved partly by selective exclusion of certain molecules and partly by active pumping of substances into or out of the cell.

The nuclear membrane

The nuclear membrane probably has general properties similar to the cell membrane; but very much less is known about it [156, 316, 508]. It presents every appearance of being a particularly tough membrane in the interphase cell and, indeed, the nucleus with its

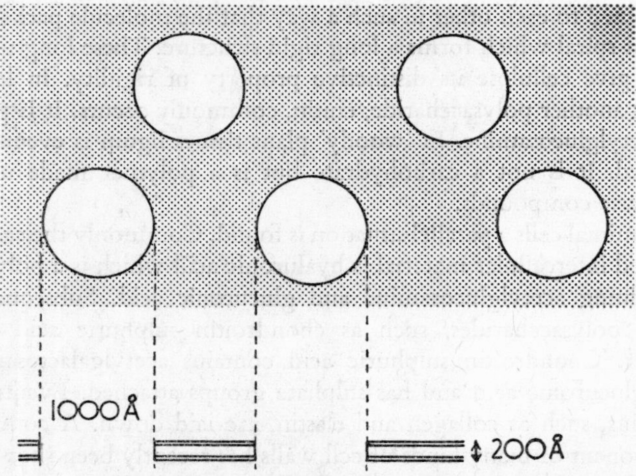

Figure **3.4. The general structure of the nuclear membrane.**

membrane often remains intact in conditions which rupture the cell membrane. Nevertheless this membrane disappears each time the cell divides. Furthermore, although it behaves as if it is impermeable to many molecules, there is excellent evidence that it contains very large pores (plate 4 and figure 3.4) [5, 13, 73, 493, 494, 513]. These annuli have a diameter of up to 1,000 Å and it has been shown in amoeba that coated gold particles of 50 to 100 Å in diameter can readily pass from the cytoplasm to the nucleus through them [136]. On the other hand conductivity measurements show that these pores are not always freely permeable and they may act as specific regulators of traffic between nucleus and cytoplasm (249).

Other membranous structures

The endoplasmic reticulum (plate 5). The reticular elements of the cytoplasm are extremely prominent in some cells. They are divided into two general types, granular (rough) membranes of the endoplasmic reticulum itself and the non-granular (smooth) membranes which are commonly associated with the Golgi apparatus. Profuse smooth membranes are also found in certain differentiated cells, particularly those concerned with the formation of steroids. The membrane of the reticulum is often directly continuous with the outer part of the nuclear membrane and is also sometimes continuous with the cell membrane itself. It is thought that it may originate from the outer part of the nuclear membrane.

Mitochondria, chloroplasts and lysosomes. These three types of intracellular organelles have skeletons of lipoprotein membranes whose structure is in each case typical. Mitochondria and chloroplasts will be discussed more fully elsewhere. Lysosomes are particles about the size of mitochondria which are rich in digestive enzymes [115, 160]. They vary in structure but are always surrounded by a lipoprotein membrane. Sometimes this is a single unit membrane, but often it consists of many layers of membrane which may be regularly coiled or irregularly folded. They are thought either to represent contracted pinocytic vacuoles or to be special digestive organelles. In damaged or dying cells they undergo lysis and release digestive enzymes into the cytoplasm. Hence, they may play a particularly important part in metamorphosis, embryonic morphogenesis and the response to injury.

Pinocytosis, Phagocytosis and Secretion. The membranes of the cell are not necessarily static. In particular, they may exhibit certain infoldings and invaginations associated with the uptake of material from the environment. Engulfment or phagocytosis is the normal manner of feeding of many lower organisms and it is observed in the scavenger cells of many higher animals. In phagocytosis a foreign particle is engulfed by the cell membrane which becomes invaginated, then forms a vacuole which is eventually pinched off.

Pinocytosis is a similar process. It was originally described in tissue cultures of animal cells but has since been shown to occur both in protozoa and in the cells of higher animals and plants in vivo [54, 223, 224, 457]. In protozoa rigid channels form in the cytoplasm. Fluid from the medium flows inwards through these channels leading to the formation of multiple small vacuoles at their bases. These minute vacuoles coalesce and move towards the centre of the cell where they are slowly absorbed.

A similar process is seen in some cells in higher animals. The brush border of some epithelial cells in the kidney tubules, for instance, consists essentially of multiple pinocytic channels very similar to the ones described in Amoeba [375]. These are probably involved in transport of substances through the cell layer. On the other hand the commonest kind of pinocytosis observed in animal cells is that seen in tissue cultures and in endothelial cells lining blood vessels [130]. In these cells pinocytosis takes the form of multiple invaginations of the cell membrane to form small vacuoles which are budded off. These coalesce into larger vacuoles which then are progressively absorbed until they eventually disappear. Some people think that these vacuoles are the origin of lysosomes [355]. The subject is discussed further in chapter 9.

Membrane flow. When vacuoles are taken into cells and ultimately disappear it is obvious that parts of the cell membrane pass into the cell interior. It has been suggested that the constituent lipoproteins may then return once more to the cell membrane thus giving rise to a flow of membrane material [33, 169]. Indeed the circulation of a particular phospholipid in this way has been proposed as a mechanism for the active transport of certain ions, notably sodium (in the salt gland of seagoing birds) [220]. It seems inherently likely that lipoprotein membranes can disrupt and reassemble readily since, as has already been mentioned, the nuclear membrane may disappear completely in the course of mitosis and reform again as soon as mitosis is completed.

Complex membranous structures and coacervates

Most intracellular structures contain lipid and protein and frequently either carbohydrate or nucleic acids also. These structures result from the orderly arrangement of the components and the question arises: to what extent do these arrangements evolve spontaneously and to what extent is their formation the result of specific regulatory processes? It can be shown that orderly structures containing lipid, protein, carbohydrate and nucleic acids may be formed spontaneously if the substances are mixed together in solution in appropriate conditions [49]. Many of these resemble cell structures, sometimes quite closely. Some structures of this type, for instance myelinic forms, have already been referred to and others will be discussed in detail later.

The complex structures called *coacervates* are of special interest. Coacervates are generally considered to arise from emulsions or macromolecular substances in solution by neutralization of the forces which normally keep the particles or molecules apart. Two of these factors are particularly important – the hydration layer round the particle and the electrostatic charge carried by it. Most colloidal biological substances form hydrophilic colloids in which the particles are surrounded by orientated water molecules. Some water may be removed from this outer layer by dehydrating agents, such as ethanol, and the particles then tend to coalesce into large aggregates, which are *simple coacervates.*

Some colloids are also maintained in solution partly as a result of their nett surface charge. If two hydrophilic colloids of opposite charge are mixed in solution they interact and form *complex coacervates.* The interaction of a basic protein, such as histone, with nucleic acids provides an example. Hence chromosomes are coacervates.

Artificial coacervates of great complexity can be formed by mixing phospholipids, nucleotides, proteins and other biological substances. Many of these resemble either intracellular structures or even cells and, as mentioned in chapter 12, it is thought that cells may have evolved from coacervates in the 'primeval broth'.

The forces which hold the components of a coacervate together are mainly of two types – electrostatic forces and Van der Waals forces (close-range absorption forces). Hydrogen bonds may contribute to strengthening the structure and occasionally covalent bonds are formed. In large particles or macromolecules electrical charges and other reactive groups may not be distributed uniformly on the surface of the molecule. Consequently, interaction between two kinds of large particles or molecules is greatly facilitated if their structures are complementary in some of these respects. If complementary molecules are present in solution in appropriate conditions they will tend to fit together like two pieces of a jigsaw puzzle. Hence no particular mechanism need be postulated for the formation of many cellular structures provided the specificity is built into the component molecules and they have an opportunity to come close enough to interact. A striking example of this type of complementarity is the quaternary structure of the haemoglobin molecule described in chapter 2. There is evidence to suggest that the assembly of some complex cytoplasmic organelles (such as the mitochondrion and chloroplast) is largely dependent on complementarity between lipoprotein molecules and some other proteins.

Part Three: The Physicochemical Basis of Cellular Activity

4. Energy in Biological Systems

Some years ago much discussion centred around the proposition that life was unique in a thermo-dynamic sense since it apparently defied the second law of thermodynamics. This law states that the entropy (that is, the randomness) of a closed system tends to increase progressively. Clearly life processes oppose the tendency to randomness and in fact are characterized by increasing order and organization. The paradox is in fact only apparent since living systems are strictly analogous to engines which release energy from fuel to do work. The entropy within an engine may not change or may even increase – but only so long as an external supply of energy is provided. A system of this kind is called an open system. Individual organisms and even the entire biosphere may be regarded as open systems.

Ordinary chemical reactions involve the making or breaking of interatomic bonds. These may be of many kinds but the commonest involve the exchange, sharing or complementation of orbital electrons in different atoms or molecules. During these reactions the energy levels of the compounds usually change and therefore energy originally present in the system is redistributed. Some may be taken up or dissipated as heat or in entropy changes.

Bioenergetics

Free energy change. When redistribution of energy occurs in any reaction the energy change conforms to the following relationship

$$(\Delta F =)\ \Delta G = \Delta H - T\Delta S$$

where Δ = change; F or G = free energy; H = heat; T = temperature; S = entropy.

This formula simply states that the total free energy change (ΔG) equals the sum of the total measurable energy change (ΔH) and the total non-measurable energy change. The measurable energy change is detectable because it can be made to do some kind of work. The non-measurable energy change involves entropy. It is usually stated that entropy is a measure of randomness, which implies that it is the inverse of order. It is sometimes easier to understand entropy in this way (that is, when ΔS is negative it implies increasing order).

The second law of thermodynamics states that all systems tend to achieve a state of randomness, that is, that they tend to spring apart into their most elementary particles. This implies that when they are held together a certain amount of potential energy is bound within them. It is the release of this energy which gives rise to an increase in entropy. It can be seen from the formula that when a reaction is accompanied by a loss of heat and an increase of entropy the value for ΔG is negative. A negative value for the free energy change means that the reaction is exergonic and may therefore proceed spontaneously, since the components of the system already contain the energy to drive the reaction. Exergonic reactions are very common. (For example, the heat loss is easily detected when fuel is burned.) Conversely, if a reaction results in an uptake of heat or a decrease in entropy ΔG is positive and the reaction is then endergonic; it requires an external source of energy to drive it.

Large increases in entropy occur when substances change from the solid to the liquid state or from liquid to gas. The latent heat involved in these changes is due to the increase in entropy. In most of the reactions encountered in biochemistry the overall entropy changes are relatively small since the reactants and products are always in solution and usually in similar concentrations. For instance, in the oxidation of glucose to water and carbon dioxide the total heat released amounts to 673,000 calories per mole while the total free energy change is 691,000 calories per mole. Hence, the entropy change amounts to a very small proportion of the total free energy change.

Activation energy. Chemical reactions are enhanced by conditions which bring molecules and atoms into close apposition so that their electronic orbits overlap and facilitate the exchange of electrons. To bring about conditions in which sufficient overlapping occurs, it is sometimes necessary first to put energy into a system (even an exergonic system) before it will proceed spontaneously. This energy is known as *activation energy* and it is exemplified in many commonplace reactions. Most substances which burn readily do not ignite spontaneously. In order to make them burn it is necessary to raise the temperature by applying extraneous heat. The effect of raised temperature is to increase the movement of electrons and atoms so that they collide frequently.

Another way to achieve the same end is to bring reactants together in very high concentrations. This can be demonstrated dramatically by freezing solutions containing reactants which do not normally combine spontaneously. As the water of the solution freezes dissolved materials are left in the unfrozen solution and reach very high concentrations, at which point they can interact [450, 460]. The free energy change involved in increasing the concentration of reactants can be calculated and is 1,420 calories per mole for every tenfold increase in concentration. This change in free energy is due almost entirely to change in entropy. It means, in simple thermodynamic terms, that this much more energy is available to drive the reactions. In subatomic terms it means that the orbits of electrons in adjacent atoms are more likely to overlap. In the long run it amounts to the same thing as putting energy into the system in the form of heat.

This type of activation is particularly important in living systems since biochemical reactions take place at a constant temperature far below that at which they would normally proceed spontaneously. It seems highly likely that many enzymes function by bringing molecules of reactants into very close apposition so that the activation conditions are satisfied.

Energy-rich (high energy) bonds. The redistribution of energy in a chemical reaction does not always lead to a large free-energy change. Instead, energy may be bound up in chemical bonds which require a particularly large amount of energy for their formation and, conversely, release a great deal of energy on being broken. Bonds of this kind are of great importance in biochemistry and they are referred to as energy-rich or high energy bonds. ('High energy bonds' are *in this sense* the opposite of the 'high energy bonds' of the chemist which are defined as bonds which require the input of a large amount of energy to break them.)

The commonest and most important kind of energy-rich bonds are the so-called high-energy phosphate bonds which appear, for example, in adenosine diphosphate (ADP) and adenosine triphosphate (ATP). These are *pyrophosphate bonds* in which one phosphate group is linked to another. On hydrolysis they yield about 8,000 calories per mole, whereas the hydrolysis of a simple orthophosphate bond releases only 2,000–3,000 calories per mole. ADP and ATP are the commonest energy carriers in biochemical systems by virtue of these bonds. Other energy-rich substances which take part in biochemical reactions are acyl phosphates, enol phosphate esters, guanidine phosphates and acyl mercaptides.

As was mentioned earlier, where reactants already contain the energy necessary to drive the reaction (ΔG is negative) then the reaction may proceed spontaneously and is called exergonic. Conversely, some reactions require an input of energy to proceed (the value of ΔG is positive) and these reactions are described as endergonic. All endergonic reactions must be linked to a source of energy. Since the reactions involved in the synthesis of biochemical macromolecules are mostly endergonic they must be linked to exergonic reactions in order to proceed. High energy compounds, usually ADP or ATP, are the immediate sources of energy for these reactions. These high energy compounds therefore act as links between exergonic and endergonic biochemical reactions and this is their main function.

The movement of electrons

Electrons are electronegative particles of very small mass. They form the outer clouds of particles which spin around the atomic nucleus and all chemical reactions occur as a result of the interchange or sharing of electrons in the outer orbits of different atoms. Because of their motion they always have energy (except in theoretical conditions).

Many important chemical reactions in living matter involve the passage of electrons (and consequently energy) from one molecule to another. If a substance acquires electrons it is said to become reduced and if it gives up electrons it is said to be oxidized. A substance which takes up electrons readily is an oxidizing agent. Oxygen is one of the most potent oxidizing agents and was the earliest to be discovered. This is why the term oxidation came to describe the more general reaction.

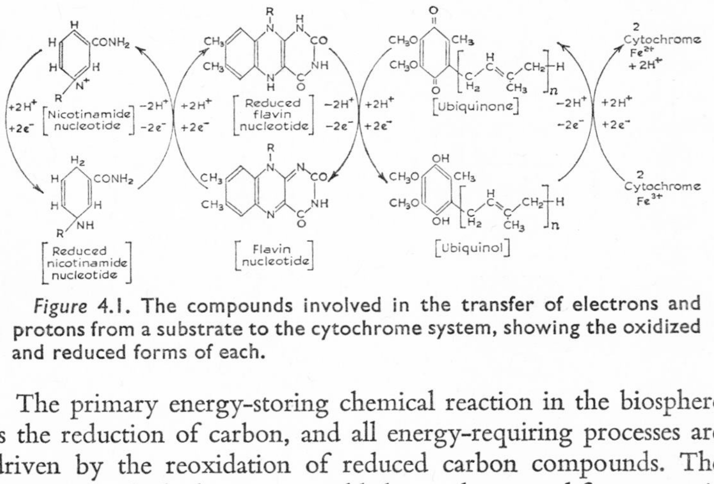

Figure 4.1. The compounds involved in the transfer of electrons and protons from a substrate to the cytochrome system, showing the oxidized and reduced forms of each.

The primary energy-storing chemical reaction in the biosphere is the reduction of carbon, and all energy-requiring processes are driven by the reoxidation of reduced carbon compounds. The manner in which electrons are added to and removed from organic substances is therefore the key to all energy metabolism within the cell. In biological oxidation-reduction (redox) systems relatively few molecules are involved. Principal among these are the quinones, nicotinamide nucleotides, flavines and metalloporphyrins [396]. Figure 4.1 shows the structure of some of the compounds in their oxidized and reduced forms. Not all these substances can react directly with molecular oxygen and some are stronger oxidizing agents than others. Consequently, they must operate in associated groups, forming sequences through which electrons can pass on their way from reduced compounds to oxygen. In living systems these electron carriers are found in association with proteins.

Oxidation-reduction potentials. Some substances give up electrons more easily than others and conversely some accept electrons more easily. It is possible to measure and assign values to these properties which are then called E_0 values.

When a metal is placed in a salt solution there is a tendency either for electrons to leave the solution and enter the metal or for electrons

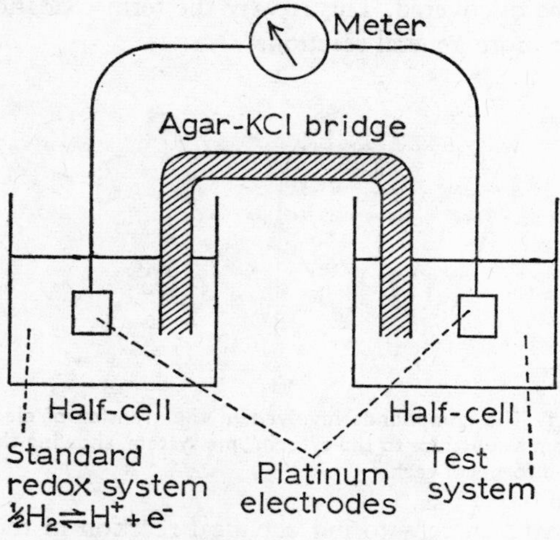

Figure 4.2. Diagram of apparatus for measuring electrode potentials.

to leave the metal and enter the solution. If two different metals in the electrochemical series are placed in the solution then there may be a tendency for electrons to leave one and enter the other. If the two are connected outside the solution by means of a wire a flow of electrons can take place, resulting in an electric current. This is the principle of the simple electric battery. If, now, instead of placing two different metals in one solution the situation is reversed and the same metal is placed in two different solutions, then the different oxidation-reduction potentials of the solutions can be compared. In practice (figure 4.2) the metal platinum is used for the electrodes, and the two solutions are placed in different vessels, connected by a bridge consisting of a high concentration of potassium chloride in a suitable non-fluid medium such as agar. This bridge permits the movement of ions but prevents reaction between the two solutions. If the two metal electrodes are connected to a potentiometer then the difference between the electrical potentials of the two solutions in the half-cells can be measured. For these measurements to be meaningful it is necessary to employ a known standard in one half-cell and to maintain standard conditions throughout. The absolute standard is the *hydrogen electrode*, in which a current of hydrogen is passed through water in which a strip of platinum foil is suspended. The hydrogen tends to dissociate into protons and electrons which tend to enter the platinum.

The *standard electrode potential*, signified by E_0', is the potential recorded for 50 per cent oxidation of an oxidation-reduction system at pH 7; it is measured in volts. When the E_0' values for a series of systems are determined they can be arranged in sequence, according to these values. In table 1 this has been done for several of the

Table 1. **Standard electrode potential values (E_0') and standard free energy changes ($\Delta G'$) in the electron transport system of mitochondria. High energy phosphate bonds are formed at the places indicated.**

Redox System	*E_0' volts*	*$\Delta E_0'$ volts*	*$\Delta G'$ per electron pair kcal/mole*
NAD $\rightarrow$ $NADH_2$	−0·3		
		+0·2	−9·2 (~P)
Flavin $\rightarrow$ Flavin H_2	−0·1		
		+0·1	−4·6
2 cytochrome bFe^{3+} $\rightarrow$ 2 cytochrome bFe^{2+}	0·0		
		+0·25	−11·6 (~P)
2 cytochrome cFe^{3+} $\rightarrow$ 2 cytochrome cFe^{2+}	+0·25		
		+0·05	−2·3
2 cytochrome aFe^{3+} $\rightarrow$ 2 cytochrome aFe^{2+}	+0·3		
		+0·5	−23·1 (~P)
$H_2O \rightarrow H_2 + O_2$	+0·8		
		+1·1	−50·8

common biological systems. In this series electrons can pass from the reduced components of one system to the oxidized components of a system with a higher E_0' value. Consequently, if all these components are placed together in solution electrons tend to flow from the members at the top of the table to those at the bottom. In the commonest type of electron transport system in living cells, this is precisely what happens as shown in figure 4.1, which describes the flow of electrons through a nicotinamide nucleotide, most commonly nicotinamide adenine dinucleotide (NAD), and along the flavine-cytochrome system.

Free energy changes in oxidation-reduction. It has been emphasized that the loss of electrons from a compound is accompanied by loss of energy. It follows that the more readily electrons are transferred in a reaction the greater is the amount of energy given up. Consequently, there is a relationship between the E_0' values and the free energy of a reaction:

$$\Delta G' = -n\, F\Delta E_0'$$

in which $\Delta G'$ is the standard free energy change, n is the number of electrons transferred per molecule and F is the faraday (equivalent to −23,063 calories per mole). These values have been inserted in table 1 and naturally they also form a sequence. It is possible to compute the free energy change involved in the transfer of electrons from one system to another by subtracting the given values from each other. It becomes clear that, whereas the transfer of electrons from hydrogen to oxygen in one single stage results in a release of 50,800 calories per mole, if the electrons pass through a series of oxidation-reduction reactions then exactly the same amount of energy is released but it is broken into smaller packages. This is what happens in biological systems. The energy released at certain stages is captured by the formation of high energy compounds in the course of this electron cascade.

Dehydrogenation. In most biochemical reactions the removal of electrons from a molecule is associated with the simultaneous removal of protons (that is, hydrogen ions). Consequently these reactions are referred to as dehydrogenations. (This term is often used in preference to oxidation since the reaction is not always completed by the combination of hydrogen with oxygen to form

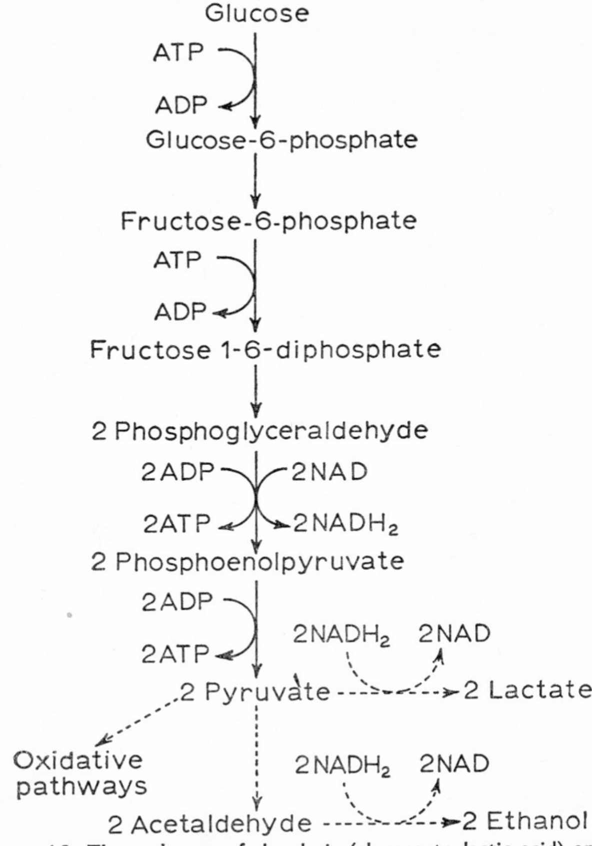

Figure **4.3. The pathways of glycolysis (glucose to lactic acid) and fermentation (glucose to ethanol) showing the points at which energy-rich compounds are formed or utilized. Note that there is a net gain of 2 ATP molecules per molecule of glucose degraded by either pathway.**

water.) The hydrogen released from a molecule during dehydrogenation has usually to be removed by combination with another substance, referred to as a hydrogen acceptor. The commonest terminal hydrogen acceptor is oxygen but many other substances can perform the function. In bacteria, for instance, sulphur and ferrous iron commonly act as hydrogen acceptors and in some organisms even nitrogen may be reduced to form ammonia and similar substances. In the simple energy-releasing pathways of alcoholic fermentation or glycolysis, acetaldehyde and pyruvic acid act as hydrogen acceptors, being reduced to alcohol and lactic acid respectively (figure 4.3).

The formation of high energy phosphate compounds [114, 274, 284, 286]. Some ATP can be formed directly in the reactions of alcoholic fermentation and glycolysis (which are thought to represent the earliest evolutionary types of energy-yielding pathways). In these

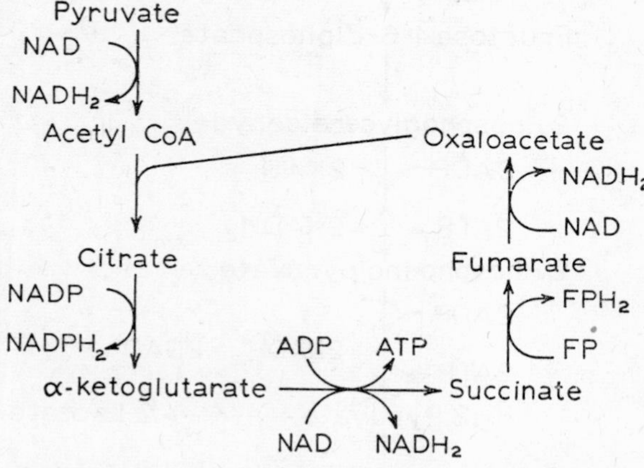

Figure 4.4. The reactions of the citric acid cycle which give rise to energy-rich compounds.

reactions, illustrated in figure 4.3, it can be seen that, in the conversion of phosphoglyceraldehyde to pyruvic acid, ADP is converted to ATP. On the other hand, two of the earlier reactions in fermentation have to be activated by high energy phosphate from ATP before they can proceed. Consequently, in the formation of four ATP molecules two are used and the net gain is only two molecules of ATP for each molecule of glucose.

The end products of these reactions, alcohol and lactic acid, are both capable of yielding considerably more energy by oxidation to carbon dioxide and water. This can be accomplished by the Krebs cycle, or variations of it, in different organisms. In this cycle (figure 4.4) the products are exposed to successive dehydrogenations. In most organisms the hydrogen released combines with oxygen to form water. However, this does not take place directly. The electrons pass along a redox system of the type described earlier (figure 4.1) which contains, in addition to a dehydrogenating enzyme (usually employing NAD as a co-factor), a system of pigments including flavines, quinones and cytochromes before oxygen is eventually reached. By this means the overall free energy change is broken up into a number of steps as shown in table 1 and figure 4.1. The table shows that in three of the steps the amount of energy released is sufficient for the synthesis of a high energy phosphate bond (about 8,000 calories). This is actually what is found to happen. One molecule of ATP is formed from a molecule of ADP at each of the three steps by the process of *oxidative phosphorylation*. Details of oxidative phosphorylation have not yet been worked out, but it has been demonstrated that an intermediate compound, with a low energy linkage, is reduced by the electrons and in the process a high energy linkage is formed which may be used to convert ADP to ATP.

The complete oxidation of carbohydrate by reactions coupled to oxidative phosphorylation is a very much more efficient energy-yielding system than glycolysis. For instance, the degradation of one molecule of glucose to two molecules of lactic acid or ethanol yields only two molecules of ATP whereas the complete oxidation of one molecule of glucose to six of carbon dioxide and six of water gives rise to no fewer than thirty-eight molecules of ATP.

The mitochondrion

The reactions of oxidative phosphorylation are associated with the mitochondrion (figure 4.5). In electron micrographs (plate 8) the mitochondrion appears as a hollow structure bounded by a double lipoprotein membrane [358, 400, 431, 437]. The inner membrane

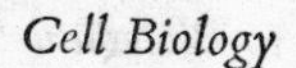

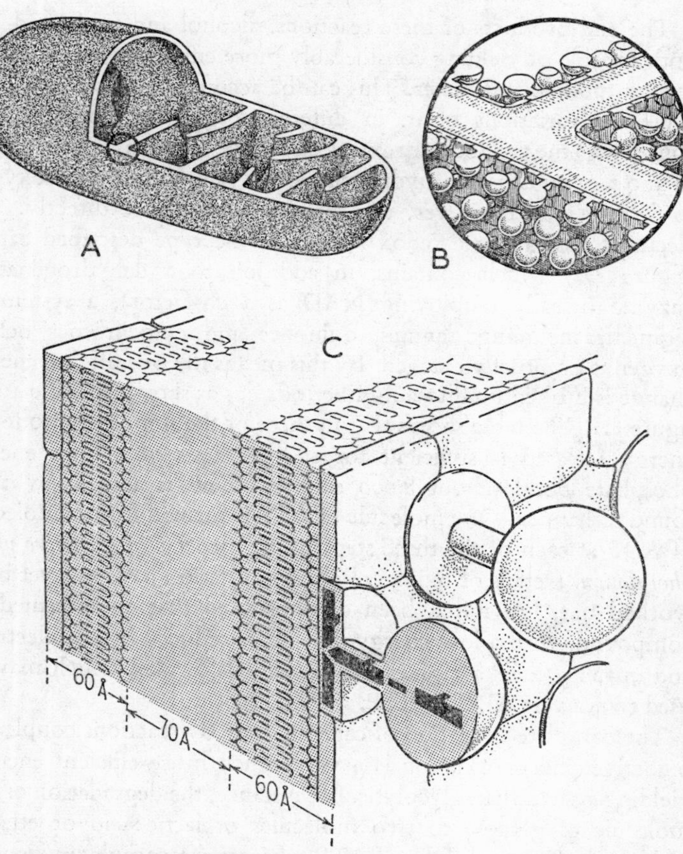

Figure 4.5. A diagrammatic representation of the structure of the mitochondrion. A. General structure, showing cristae. B. Section of a crista showing how elementary particles may be disposed on the surface. C. The molecular skeleton of the mitochondrion.

is convoluted to form the *mitochondrial cristae*. Mitochondria may vary considerably in shape from spherical to much elongated. They contain the cytochromes, the dehydrogenase enzymes associated with them, the flavine respiratory pigments and some enzymes involved in the Krebs cycle and lipid metabolism. These do not lie free within the mitochondrion but form an association with the lipid membrane on which they are arranged in a specific order (figure 4.6).

From mitochondria it is possible to extract particles (elementary particles) which have a molecular weight of less than a million [178]. These can be broken into four complexes (figure 4.6). Two are dehydrogenase complexes, of which one reduces succinic acid and the other is a general reducing complex using NAD as a co-factor. The other two constitute the so-called electron transport system (ETS). The ETS can be separated into two constituent complexes, one of which contains the enzymes and pigments which catalyse the pathway from ubiquinone to cytochrome C and the other the transport of electrons from cytochrome C to oxygen. Each has a molecular weight of less than 200,000. If they are broken down

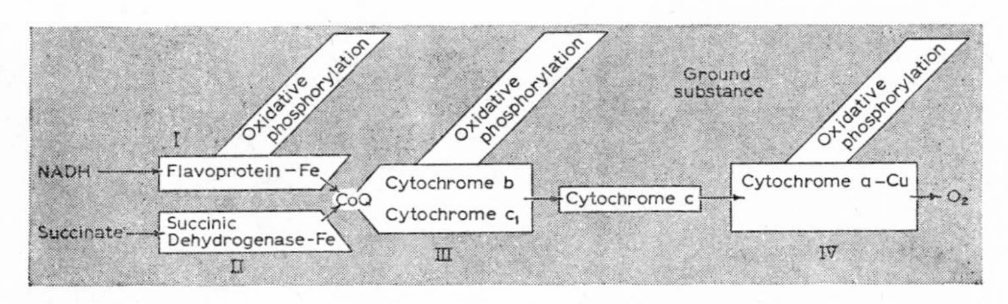

Figure 4.6. The four parts of the elementary particle showing diagrammatically how they are probably arranged to facilitate electron transport and oxidative phosphorylation.

further they cease to be active but they can be recombined and then regain their activity.

The complexes can be combined in groups [99, 151, 179, 201, 528]. Together they can form the kind of arrangement shown in figure 4.6. Certain enzymes in the mitochondrion require lipids as co-factors. One in particular, isobutyric dehydrogenase, has a specific requirement for lecithin. Since it is most unlikely that this substance participates directly in the transfer of electrons it probably acts by stabilizing the tertiary structure of the enzyme so that its configuration is suitable for the performance of its catalytic function. Recently evidence has been obtained from electron microscopic studies that the surface of the mitochondrial cristae is studded with small spheres (plate 9) [141, 455]. These are about the right size for elementary particles and there is some evidence that they are associated with electron transport activity. Hence, the mitochondrion probably consists of a lipoprotein scaffolding on which are arranged elementary particles and enzymes involved in associated reactions (figure 4.5). It seems very likely that they are distributed in an orderly array, like an assembly line, so that products of one reaction can proceed directly to the next enzyme in a series.

5. Energy Transducers

An energy transducer is a device which converts one kind of energy into another. Transducers are of great importance in biological processes and several kinds can be distinguished.

Radiant to chemical energy; the interaction of radiant energy and matter

Radiant energy can be considered in two ways, either as a wave motion or as a series of small particles. When radiant energy is absorbed by atoms it causes movement of subatomic particles, mainly electrons. High energy radiations, such as X-rays, may displace electrons entirely, giving rise to ionized atoms or molecules which are highly chemically reactive. Visible light, infra-red and ultraviolet radiations carry relatively lower energy and they usually succeed only in moving electrons into orbits of higher energy. This may also result in atoms and molecules becoming more chemically reactive since it amounts to introducing activation energy.

When an electron is moved into a higher energy orbit as a result of being struck by a photon (quantum of radiant energy) it may either fall back again into a lower energy orbit almost immediately and emit a photon in so doing, or if it is moved far enough, it may interact with the outer orbit of a neighbouring atom. In the former event fluorescence results; in the second a chemical reaction takes place.

It is easy to appreciate how atoms can be activated in this way. In large molecules, however, not every atom may be in a position to react with a neighbouring molecule and, consequently, much of the incident radiant energy is liable to be wasted. In some molecules however a wave of electronic excitation can be set up. A photon may strike a molecule at a part distant from the reactive region and excite an electron in that area. When this electron falls to the ground state its energy may be transmitted to an electron in a neighbouring atom; in this way the excitation may be propagated through the molecule until an electron at the periphery is raised to a sufficiently high energy level to react with a neighbouring atom. The wave of excitation requires no actual movement of electrons from one atom to another. It is described as a movement of excitons and is analogous to wave motion [260, 399].

This mechanism can greatly increase the efficiency of radiation in activating chemical reactions between molecules. The higher the energy of the radiation (that is, the shorter the wave length) the more effective it is in promoting reactions. The abiogenic synthesis of organic molecules in early evolution was probably stimulated directly by short wave ultraviolet radiation.

Photosynthesis

In photosynthesis light energy is used to drive biosynthetic reactions [17, 213]. Impingement of photons initiates a movement of excitons which may eventually result in an electron being raised into an orbit where it can be captured by an adjacent electron acceptor. When this happens a movement of electrons may occur in the entire molecule in an attempt to fill up the 'hole'. If an electron donor happens to be in close apposition to any part of the molecule it may give up an electron to fill the hole, and the final effect may therefore be to cause an electron to move from an electron-donating substance to an electron-accepting substance. Radiant energy absorbed during this process may thus succeed in driving an oxidation-reduction reaction which would not occur spontaneously because the free energy change would be too large. This is the manner in which light energy is captured by chlorophyll and used for the photolysis of water [15, 16, 76].

In photophosphorylation (figures 5.1, 5.2 and 5.3) chlorophyll is the photosensitive material. In cyclic photophosphorylation (figure

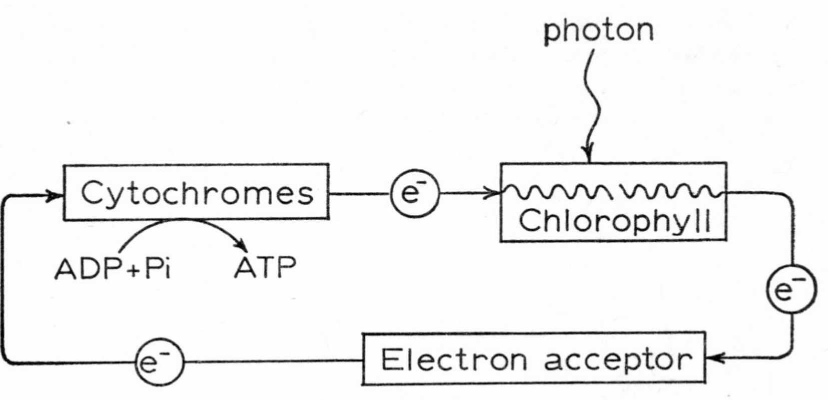

Figure 5.1. The central reactions of cyclic photophosphorylation.

5.1) the electrons, displaced from chlorophyll into an electron acceptor, return to chlorophyll by a series of reactions which amount to an electron cascade. Some of their energy is captured for use in other reactions. In non-cyclic photophosphorylation an independent electron donor is required. In bacteria this may be a substance such as thiosulphate or succinate (figure 5.2).

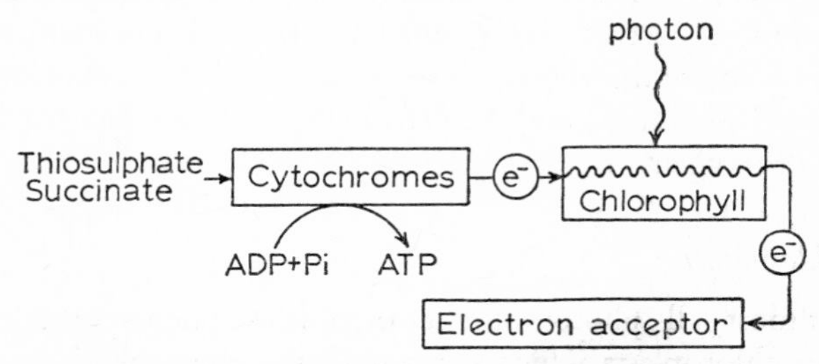

Figure 5.2. The central reactions of non-cyclic photophosphorylation, of the type observed in bacteria.

In the common type of photophosphorylation seen in green plants (figure 5.3) one of the electron acceptors (probably not the immediate one) is the co-enzyme nicotinamide adenine dinucleotide phosphate (NADP). Each molecule of NADP can take up two electrons to form $NADPH_2$ which can then be used for synthetic processes. Two protons are also required and these are obtained by the breakdown of water. The other products of this breakdown are hydroxyl ions and electrons. The hydroxyl ions react to form oxygen and water while replacement of the electrons in chloro-

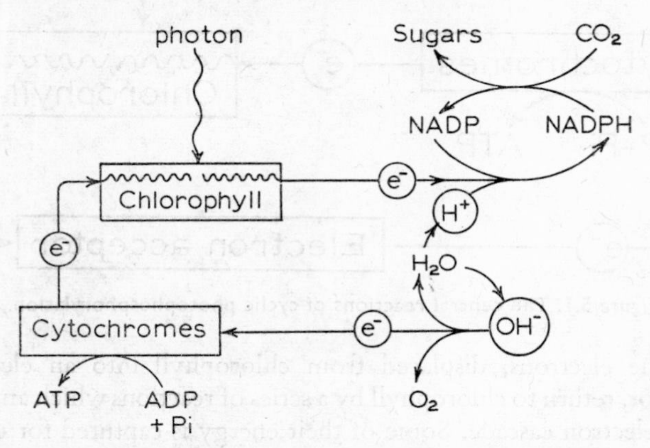

Figure **5.3. The central reactions of non-cyclic photophosphorylation, of the type commonly found in green plants.**

phyll occurs by a flow of electrons (from water) through a chain of cytochrome pigments, during which ATP is formed from ADP. The end result of the photolysis of water is, therefore, the formation of reduced $NADPH_2$ and ATP from NADP and ADP, with the release of oxygen.

The chloroplast

In almost all photosynthetic organisms photosynthesis takes place in chloroplasts which contain all the chlorophyll of cells in which they are present. With the electron microscope (plate 10) the chloroplast is seen in section to consist of parallel lamellae of lipoprotein [216]. In most species there are areas of increased lamellar density called grana which are formed by the interpolation of additional lamellae containing chlorophyll. In disrupted chloroplasts the grana can be shown by shadowing techniques to consist of stacks of discs, the surfaces of which carry a regular repeating

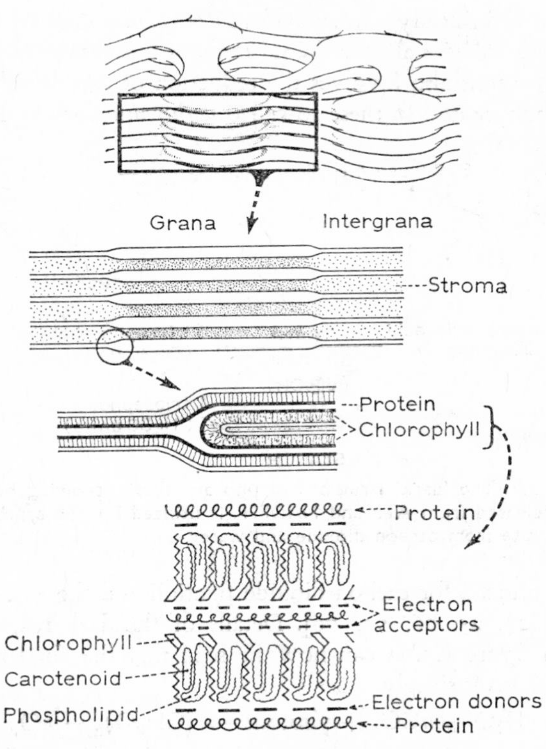

Figure **5.4. The molecular structure postulated for the chloroplast, showing how a directional flow of electrons may be achieved when photons are absorbed by chlorophyll.**

array of substructures, called quantasomes [362]. It is within the grana that the *light reactions* of photosynthesis occur, that is, the reactions involving photolysis of water and resulting in the synthesis of $NADPH_2$ and ATP. It is thought that the grana are composed of layers of oriented chlorophyll molecules sandwiched between layers of electron donors and electron acceptors (figure 5.4) to form a kind of solar battery in which the impingement of photons results in a steady flow of electrons from donors to acceptors and consequently in steady synthesis of ATP. Quantasomes are thought to represent individual functioning units of this system.

Photosynthesis also involves the reduction of carbon dioxide to form carbohydrates. In these so-called *dark reactions* $NADPH_2$ and

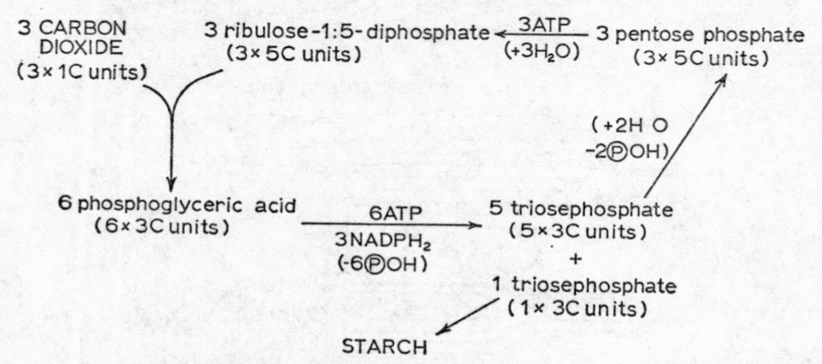

Figure 5.5. The 'dark' reactions of photosynthesis, showing how the energy released from ATP and $NADPH_2$ is utilized for the synthesis of carbohydrate from carbon dioxide and water.

ATP are utilized for carbon dioxide fixation and the end-product is starch [24, 25, 26, 75, 76]. Details of the dark reactions are shown in figure 5.5. It can now be seen that the final result of photosynthesis is simply

$$nH_2O + nCO_2 + \text{photons} \rightarrow (CH_2O)_n + nO_2$$

Chloroplasts belong to a group of intracellular organelles called plastids, some of which are colourless and cannot perform photolysis, although they can perform the dark reaction (the synthesis of starch), provided $NADPH_2$ and ATP are available from some other source.

Light receptors

A similar but less well understood mechanism to that described above is involved in vision. The photo-sensitive substances are also pigments and there are probably several of them, the best-known being visual purple derived from vitamin A. The visual pigments occur in the cells of the retina where they are associated with laminated structures which bear a superficial resemblance to the grana of chloroplasts [319, 326, 430]. It seems that mechanisms similar to the light reactions of photosynthesis are involved in the functioning of these structures. The details are not well understood, but the primary reaction is probably a switch of vitamin A aldehyde from the *cis* to the *trans* form on the absorption of photons.

Chemical to radiant energy

The opposite type of transducer also exists – the result in this case being the conversion of chemical energy to light energy in the form of luminescence or phosphorescence [299, 300].

When electrons in an atom are raised to a higher energy orbit and then fall back to their original orbit photons may be emitted. This is the basis of fluorescence. There are many different kinds of bioluminescent systems, of which the essential component is a 'luciferin', a substance which emits photons when altered in a chemical reaction. Alteration of the compound usually depends on an enzyme or series of enzymes called luciferase. ATP or a dehydrogenase system is usually required to provide energy to raise electrons to a higher orbit. One of the best known systems is the one found in fireflies (Lampyridae), in which ATP reacts with luciferin in the presence of luciferase to produce fluorescence. The luciferin has been identified as 2-(4′-carboxythiazolyl)-6-hydroxybenzothiazole. Another system in bacteria utilizes as luciferin an aldehyde complex of reduced riboflavin-5′ phosphate, which emits light when it is oxidized.

Chemical to mechanical energy

Contractility is one of the most important properties of all living cells. The precise mechanisms involved in the contraction of cytoplasm are not completely understood but it is known that ATP is nearly always broken down to ADP. Indeed this part of the process can be demonstrated in glycerol extracted muscle tissue which will contract if ATP is added to the medium [52, 459]. A similar phenomenon can be observed with glycerol-extracted flagellated or ciliated organisms in which the flagella or cilia begin to beat rhythmically when ATP is added [166, 219]. Glycerol-extracted amoebae and fibroblasts also show movements on addition of ATP [11, 204, 218, 427].

Many mechanisms have been proposed to explain the role of ATP in contraction. The polyelectrolyte system exemplifies one possibility (figure 5.6) [50, 327]. This proposes that, in the absence

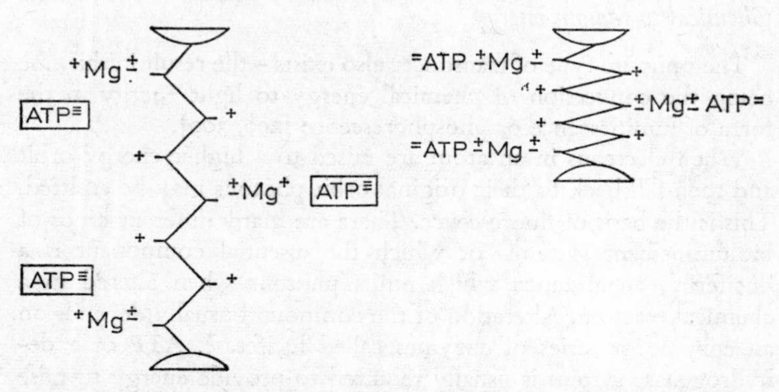

Figure **5.6. The polyelectrolyte model of protein contraction.**

of other forces, the contractile molecule has a tendency to collapse and shorten owing to straining of bonds. Collapse is, however, prevented because similar positive charges are distributed along the surface of the molecule, and these repel each other. It is suggested that ATP acts by neutralizing the positive charges on the molecule and permitting it to collapse to the shortened form.

Another theory proposes that the molecule is kept extended by hydration and that disturbance of the ionic balance causes the molecule to shrink into a supercontracted form [85]. Models of these kinds have been proposed to explain the contraction of many proteins, including muscle protein. However, an entirely different principle has been proposed for muscle by Huxley and Hansen on the basis of its known ultrastructure [195, 215, 237, 239, 435].

When viewed with the electron microscope (plates 12, 13) muscle in longitudinal section is seen to consist of alternating bands [51].

The wide dark band is known as the *A* band and the narrow one the *Z* band (figure 5.7). The clear area between the *A* and *Z* band is called the *I* band. When muscle contracts the *H* band disappears and the *Z* and the *A* bands are brought into close apposition. At

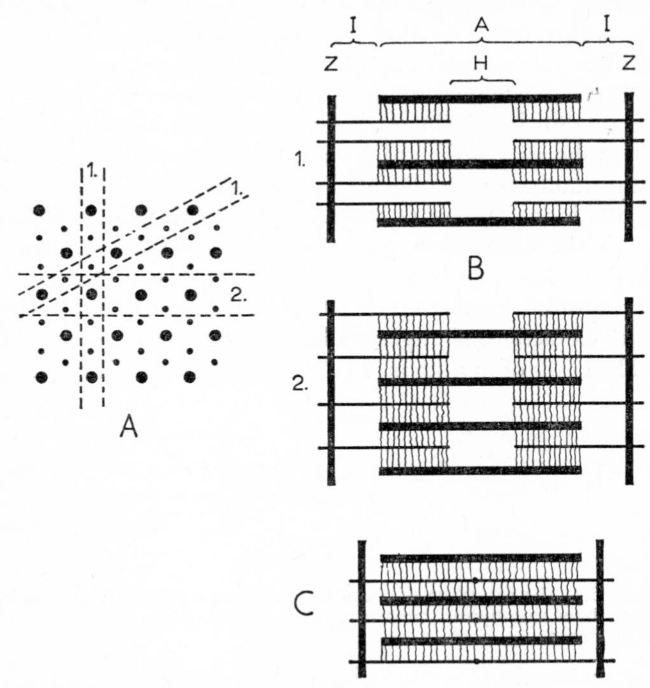

Figure 5.7. Diagrammatic representation of the ultrastructure of muscle. Depending on whether a longitudinal section passes through the planes 1 or 2 in A (which represents a cross-section) the appearances shown in B1 or 2 are obtained with the electron microscope, in resting muscle. In contracted muscle the H-band disappears and the *Z* bands are pulled closer together as shown in C.

high magnifications (plate 13B) it can be seen that fine fibrils, about 1 μ in length and 50 Å in diameter, extend from the *Z* bands into the *A* band area where they interdigitate with more heavily stained fibres, about 1·5 μ in length and 150 Å in diameter. The thick fibres characteristic of the *A* band consist of the protein myosin whereas the thin fibres extending from the *Z* bands consist of the protein actin. Actin and myosin can be extracted together from muscle as actomyosin, and Szent-Gyorgi showed that actomyosin fibres isolated in this way will contract on the addition of magnesium and ATP to the surrounding medium [394]. During this contraction ATP is broken down to ADP and inorganic phosphate. The actomyosin fibres do not relax again unless an ATP-generating system (creatine phosphate plus creatine phosphate kinase) is added.

Pure myosin has ATPase activity [247, 294). Moreover, it has now been shown that the 150 Å diameter myosin fibre is composed of a large number of interdigitated molecules, each of which is composed of a long filament (1,500 Å) with a knob-like head (figure 5.8). It is the filamentous parts which interdigitate to form the 150 Å fibre while the knob-like heads (which have the ATPase activity) protrude on the surface where they form the cross-bridges between the actin and myosin filaments seen in plate 13B. The knob-like heads of myosin molecules interact in a spatially oriented

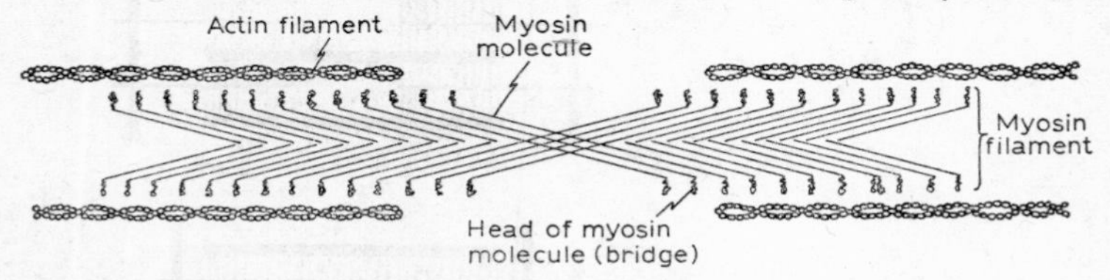

Figure **5.8. The structure of the myosin and actin filaments of muscle, showing how the myosin molecules interact with each other and with actin filaments.**

manner with actin fibres (which are themselves made up of large numbers of globular actin molecules). These facts have led to the belief that an interaction between myosin and actin, utilizing ATP, accounts for muscle contraction and this would seem to fit in with the electron microscopic ultrastructure shown in plate 13B. Huxley and Hansen have proposed a mechanism whereby an actin fibre slides along a myosin fibre, possibly by the alternate making and breaking of bond connections between the two [238, 240].

Cilia and flagella. There is good evidence that these two motile structures also use ATP since both of them have been shown to be very rich in ATPase. Cilia and flagella are distinguished by a unique ultrastructure which is found at all levels of evolution. Within each individual cilium there are always nine pairs of fibrils which usually surround a core consisting of two large fibrils (plate 14A) [18, 55, 129, 134, 135, 305, 440]. ATPase activity has been demonstrated in association with the outer ring of fibrils by histochemistry. Probably the contractile molecules act by pulling on the fibrils.

In ciliated organisms the bases of the cilia may be joined by a structure called the kinetie. In organisms such as *Paramecium* this forms a kind of exoskeleton which is highly orientated and which has been shown to play an important part in the morphogenesis of this single-celled organism.

The 'nine plus two' structure also turns up in the centrioles, although here the central two fibres are missing and only the outer nine groups persist (plate 14B). In some cases it seems that when flagella develop they arise from the centrioles. The significance of this pattern is not yet understood but it almost always seems to be associated with motile or contractile structures and ATPase activity.

Cytoplasmic movement. Within the cytoplasm many random movements occur which can be explained simply on the basis of thermal and diffusional forces. However, in addition, cells exhibit many purposive movements such as those involved in locomotion and division.

Three kinds of movement seem to be distinguishable in cellular locomotion. (i) Some cells move by virtue of cilia, which propel them over a surface in much the same manner as a millipede. (ii) Other cells exhibit a snail-like pattern of locomotion which is produced by the movement of alternating waves of compression and relaxation along the membrane of the cell, giving rise to a typical undulating motion. (iii) A third pattern, exhibited particularly by amoebae, involves the continuous forward flow of cytoplasm.

The mechanisms involved in the two latter modes of locomotion are not understood. Suggestions have included contraction of elements in the cytoplasm, a direct flow of cytoplasmic contents and sol-gel transformation in the cytoplasm. All of these may function in different instances but there is increasing evidence that most locomotory behaviour is associated with contractile fibrils in the cytoplasm of which the fibrils of cilia and flagella may represent specialized forms. Hence the contractile movements of cytoplasm involved in cytoplasmic movement may be closely related to other contractile phenomena.

Chemical to osmotic energy

In all cells the internal environment differs in many important respects from the external environment. In particular certain ions are either concentrated within the cytoplasm or excluded from it [64, 67, 196, 480]. This is achieved by the so-called ionic pumps. The most thoroughly studied of these is the sodium pump [70, 167, 197, 411, 507], which occurs in all animal cells but there is also

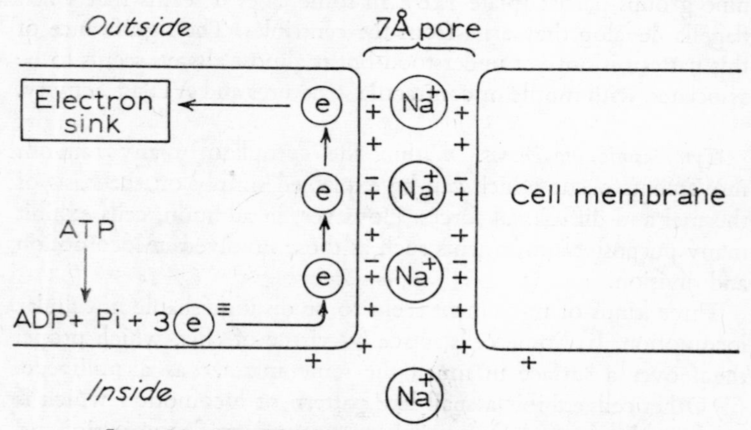

Figure 5.9. A model for the sodium pump, postulated by Skou.

evidence for potassium pumps in plant cells and for chloride pumps in other cells.

The sodium pump is closely associated with the plasma membrane of the cell. In the removal of sodium ions ATP is broken down to ADP and inorganic phosphate. The process requires potassium and magnesium. On the average three sodium ions are pumped out of the cell for each ATP molecule broken down. The possible nature of the sodium pump has been much illuminated by the isolation of a membrane-bound ATPase which is dependent on sodium, potassium and magnesium ions [507]. This ATPase operates asymmetrically, sodium ions appearing at one side of the membrane and potassium ions at the other. Several mechanisms have been proposed to explain how this sodium pump works but perhaps the most ingenious is the one put forward by Skou (figure 5.9). He suggests that the sodium pump is located at the pores in the cell membrane which are lined with positive charges which prevent the passage of sodium ions. He postulates that when ATP is broken down to ADP the three electrons released appear at the intracellular end of the pore where they make the electrical field momentarily more negative. The electrons are then imagined to travel outwards along the pore carrying a wave of electro-negativity with them and permitting the passage of a sodium ion alongside each electron. Potassium ions, being smaller than sodium ions in the hydrated form, may pass in through the middle of the pores to restore the ionic balance.

Chemical to electrical energy [402]

When electrolytes are maintained at different concentrations on either side of semi-permeable membranes electrical potentials are set up; these can be demonstrated experimentally across cell membranes by inserting a microelectrode. The semi-permeable properties of biological membranes are apparently easily disturbed. If a stimulus is applied to a membrane it may be momentarily depolarized and the state of depolarization can be propagated, carrying with it a momentary change in electrical potential. This results in the movement of an electrical charge across the surface of the cell.

This behaviour is a property of very many cells but is, of course, particularly highly developed in the nerve cell. The depolarization is only momentary since the ionic pumps restore the electrolyte balance to its original state in a matter of milliseconds, using ATP in the process.

6. Synthesis of Proteins and Nucleic Acids

The intimate relationship between proteins and nucleic acids has already been mentioned. The most important feature of the relationship is that DNA is considered to be the 'genetic material' which, by means of a code formed by the orderly sequence of its constituent nucleotides, determines the order of amino acids in proteins (and therefore their properties). RNA appears to operate as an intermediary between DNA and proteins.

DNA as genetic material

It is now almost conclusively proved that genes are, in fact, made of DNA [230]. The evidence is as follows:

(i) DNA is metabolically stable and is passed to daughter cells intact.
(ii) DNA is intimately associated with chromosomes, and the amount of DNA per cell is related to the chromosomal content of the cell.
(iii) Chemical and physical factors known to alter the structure of DNA act as mutagens and result in specific changes in proteins.
(iv) Genetic information can be introduced into cells by means of pure DNA (transformation and transduction).

Stability of DNA. It was first shown by labelling the DNA of bacteriophage (viruses which attack bacteria) with radioactive phosphorus that a block of DNA was always passed on intact to the progeny [208, 209, 211, 292]. Subsequently, it was shown in bacteria and animal cells that, if DNA was labelled with an isotope and the cells were then grown for many generations in the absence of the isotope, none of the isotope was lost [428, 472]. These experiments indicated that the DNA molecule was not broken down during cell division. However, this kind of evidence is not conclusive since it is possible for the breakdown products of DNA to be immediately reincorporated into new DNA to give the same result.

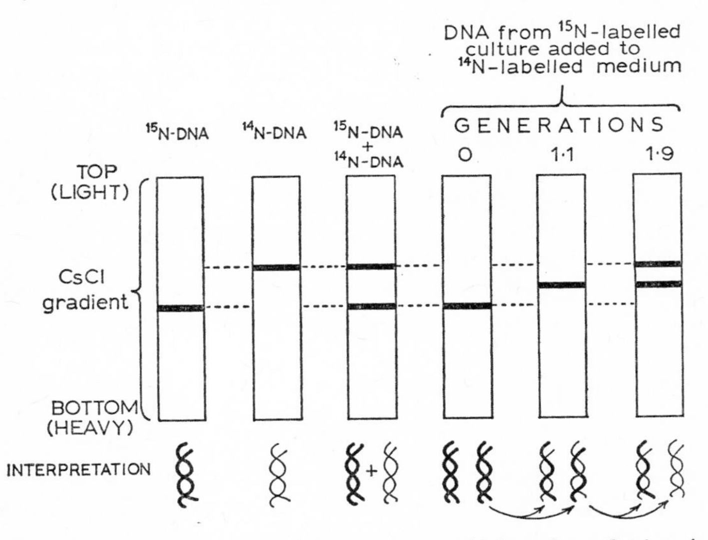

Figure **6.1.** **Meselson and Stahl's experiment which confirms Crick and Watson's theory for the replication of DNA. Escherichia coli were grown in a medium containing heavy nitrogen (^{15}N) and then transferred to a medium containing ordinary nitrogen (^{14}N). As shown in the three figures at the left DNA containing ^{15}N can be distinguished from DNA containing ^{14}N by centrifugation at very high speeds in a gradient of cesium chloride. As shown in the three figures at the right an intermediary band appeared in bacteria grown in the above conditions, indicating a 'semiconservative' mode of replication of the molecule.**

An experiment by Meselson and Stahl (figure 6.1) provided a less equivocal answer to this question [317]. Bacteria were grown for several generations in a medium containing heavy nitrogen, ^{15}N, instead of the normal isotope, ^{14}N. All DNA synthesized during this time was heavier than DNA formed in ordinary conditions. The bacteria were then permitted to continue growing in an ordinary medium (containing only ^{14}N) and samples were periodically removed. DNA was isolated from these and placed in a density gradient, made of a cesium chloride solution, graduating from a very dense concentration at the bottom of the tube to a more dilute one at the top. The specimens were centrifuged at very high speed until the DNA settled at an equilibrium position, depending on its density relative to that of the gradient. As shown in the diagram it was easily possible to distinguish between ^{15}N-DNA and ^{14}N-DNA. At first almost all the bacterial DNA was heavy (^{15}N). Then a DNA of intermediate density appeared at about the time when most of the cells would have completed their first division, indicating the formation of a mixed or hybrid molecule. At the next generation the intermediate band persisted but a light band also made its appearance. If this course of events is compared with the Watson-Crick model of DNA replication (page 27) it will be seen that it is completely compatible with it. Other experiments of this type have been done and they support this conclusion.

Association of DNA with chromosomes. Some specific histochemical tests for DNA have been evolved, the best known of which is the Feulgen stain. In cells with recognizable chromosomes the staining for DNA is confined almost entirely to them. These observations indicate that DNA is located in the chromosomes which have been proved by genetic means to carry the genes. Furthermore, haploid cells, such as spermatozoa, which have half the number of chromosomes of diploid somatic cells, also have half as much DNA. Cells with more than the normal number of chromosomes (such as tetraploid cells) have proportionally more DNA [44, 482]. Moreover, when radioactive thymine (which is incorporated only into DNA) is fed to cells and these are later covered with a photographic emulsion, the silver grains arising in the emulsion as a result of radioactive decay are situated almost entirely over the chromosomes [72].

Action of Mutagens. Many mutagens, such as irradiation, nitrous acid, hydroxylamine and bromo-uracil, produce chemical changes in DNA and these result in changes in some proteins. For instance, nitrous acid deaminates guanine, cytosine and adenine to give xanthine, uracil and hypoxanthine respectively [152]. Mutations produced by this substance are distinguished by substitution of one amino acid for another in proteins and these changes fit into the scheme for the genetic code which will be discussed later (page 132)

Transformation and transduction. If a mutant bacterium, which has lost a gene, is treated with pure DNA from the wild-type organism

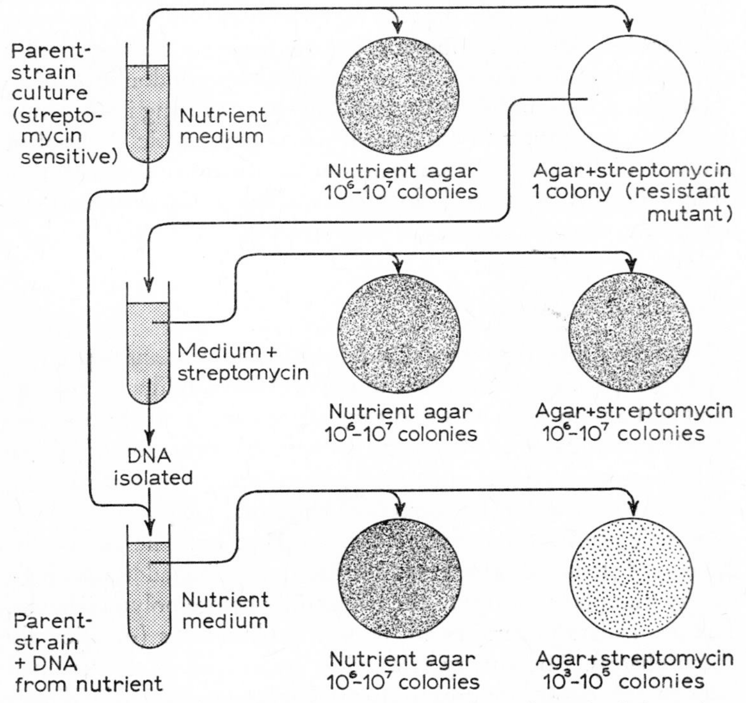

Figure **6.2. The experimental demonstration of transformation in bacteria. Streptomycin is lethal to wild-type organisms but resistant mutants arise rarely. These can be isolated by growing the bacteria in the presence of streptomycin. If DNA is prepared from a resistant strain and added to a culture of wild-type organisms many more streptomycin-resistant bacteria appear than would be expected by mutation. It is concluded that DNA entered some of the wild-type bacteria and give rise to the resistance.**

containing the gene then the lost activity can sometimes be restored [21, 181]. The organism then behaves like the wild-type and breeds true. Hence the introduction of DNA to the organism is equivalent to introduction of a gene. The process, called *transformation*, is illustrated in figure 6.2. In *transduction* a similar phenomenon occurs but in this instance a bacteriophage acts as a natural carrier of the gene [529]. This is discussed later in connection with bacterial genetics.

None of these pieces of evidence by itself is completely conclusive but, taken together, they provide very strong evidence for the view that DNA is the most important genetic material. So many other experiments lead to the same conclusion that no real doubt remains.

DNA is not the only substance capable of conveying information from one generation to another. RNA is the genetic material of many viruses while there are well-known instances of information being inherited by quite different means, but DNA performs this function in the vast majority of organisms.

Synthesis of nucleic acids

Since the properties of both nucleic acids and proteins depend on the sequence of the component monomers (nucleotides on the one hand and amino acids on the other), the synthesis of each resolves itself into two parts, one having to do with the formation of chemical linkages between the monomers and the other with the ordering of the components in the correct sequence [81, 111].

Nucleic acids are synthesized from nucleotides. The energy for the phosphate diester linkages is obtained by adding additional high energy phosphate groups to the nucleotides before polymerization. The immediate precursors are usually triphosphates (occasionally diphosphates). The energy released in hydrolysis of the terminal high energy phosphate groups is partly incorporated in the diester linkage. Details of the process vary and must be considered in relation to each individual nucleic acid.

Deoxyribonucleic acid [45, 272, 273, 283, 304]. In preparation for the synthesis of DNA the deoxyribonucleoside monophos-

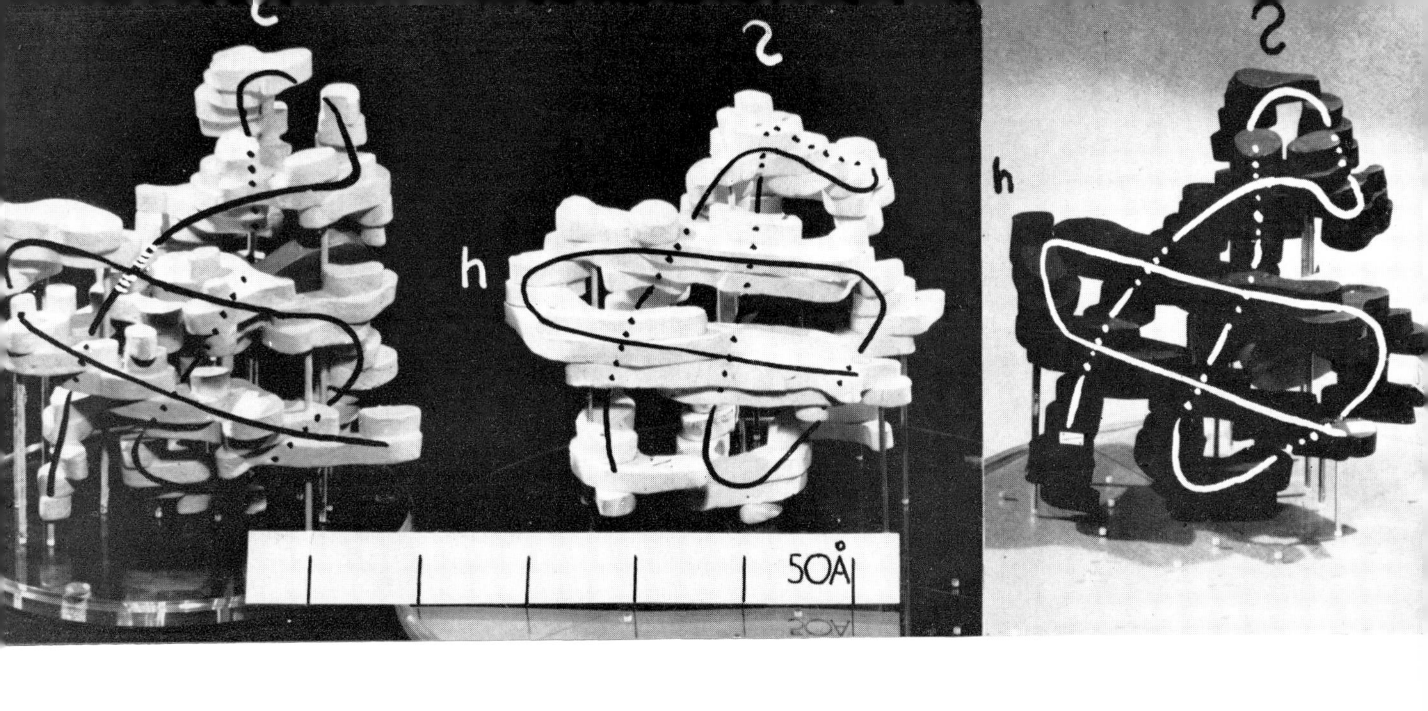

Plate 1. From left to right. Models of the molecule of myoglobin and the α- and β-chains of haemoglobin. (*Courtesy of Dr M. F. Perutz*)

Plate 2. The association of two α-chains and β-chains to form the complete molecule of haemoglobin. (*Courtesy of Dr M. F. Perutz*)

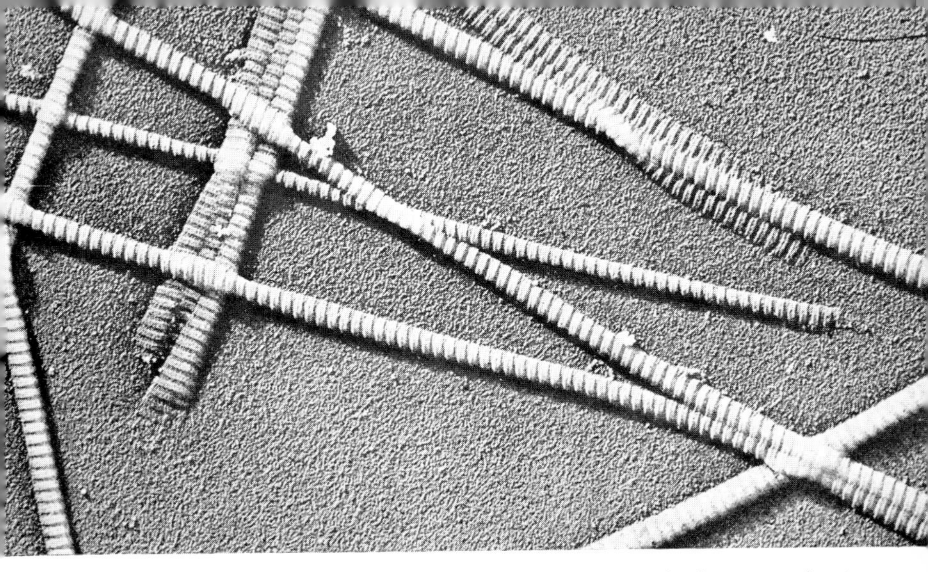

Plate 3A. Electron micrograph of collagen, shadowed with chromium, showing bands repeating with a period of 640 Å (×25,000). (*Courtesy of Dr J. Gross*)

Plate 3B. Electron micrograph of sciatic nerve with (*inset*) X-ray diffraction diagram of the same material (osmium fixed). (*Courtesy of Dr H. Fernandez Moran*)

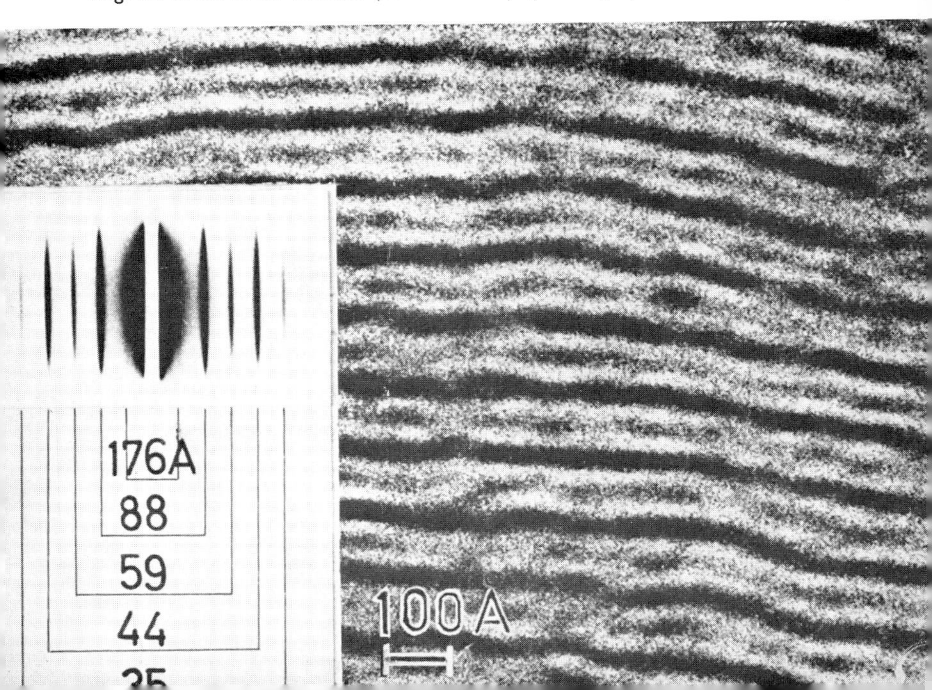

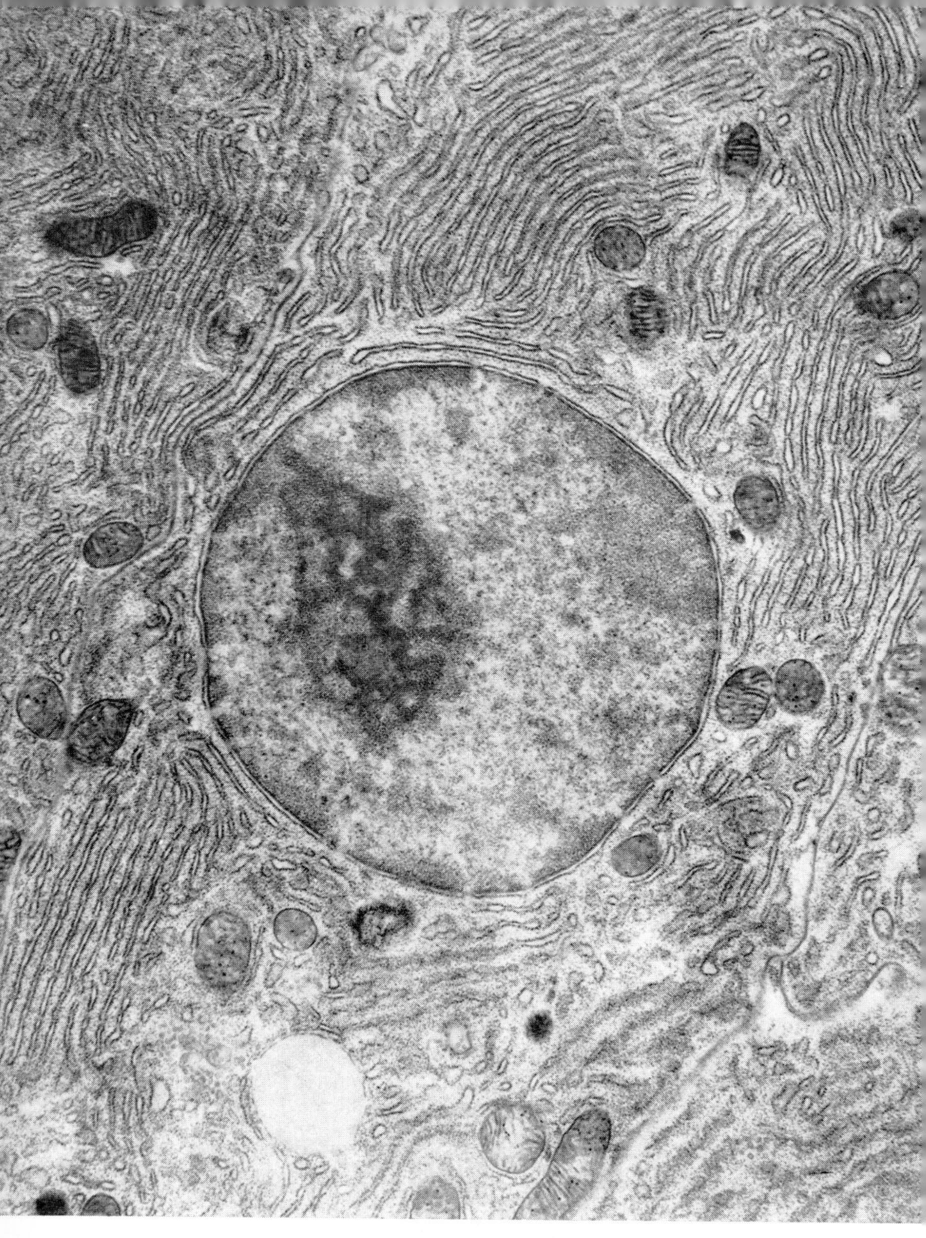

Plate 4. Electron micrograph of a pancreatic acinar cell showing the nuclear membrane, with prominent annuli, and the nucleolus (× 18,000). (*Courtesy of Dr D. W. Fawcett*)

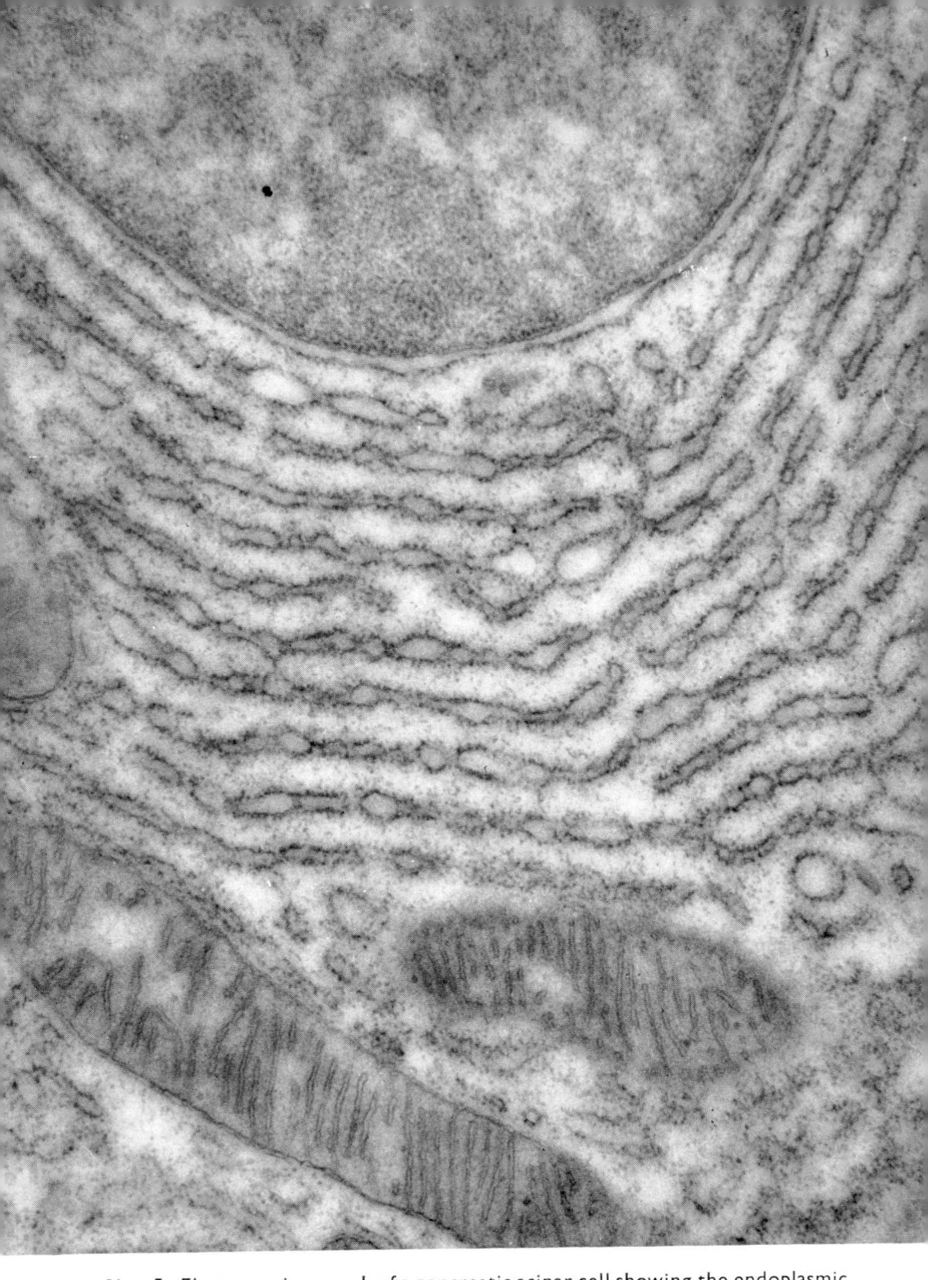

Plate 5. Electron micrograph of a pancreatic acinar cell showing the endoplasmic reticulum, free ribosomes, and several mitochondria (×36,000). (*Courtesy of Dr G. Palade*)

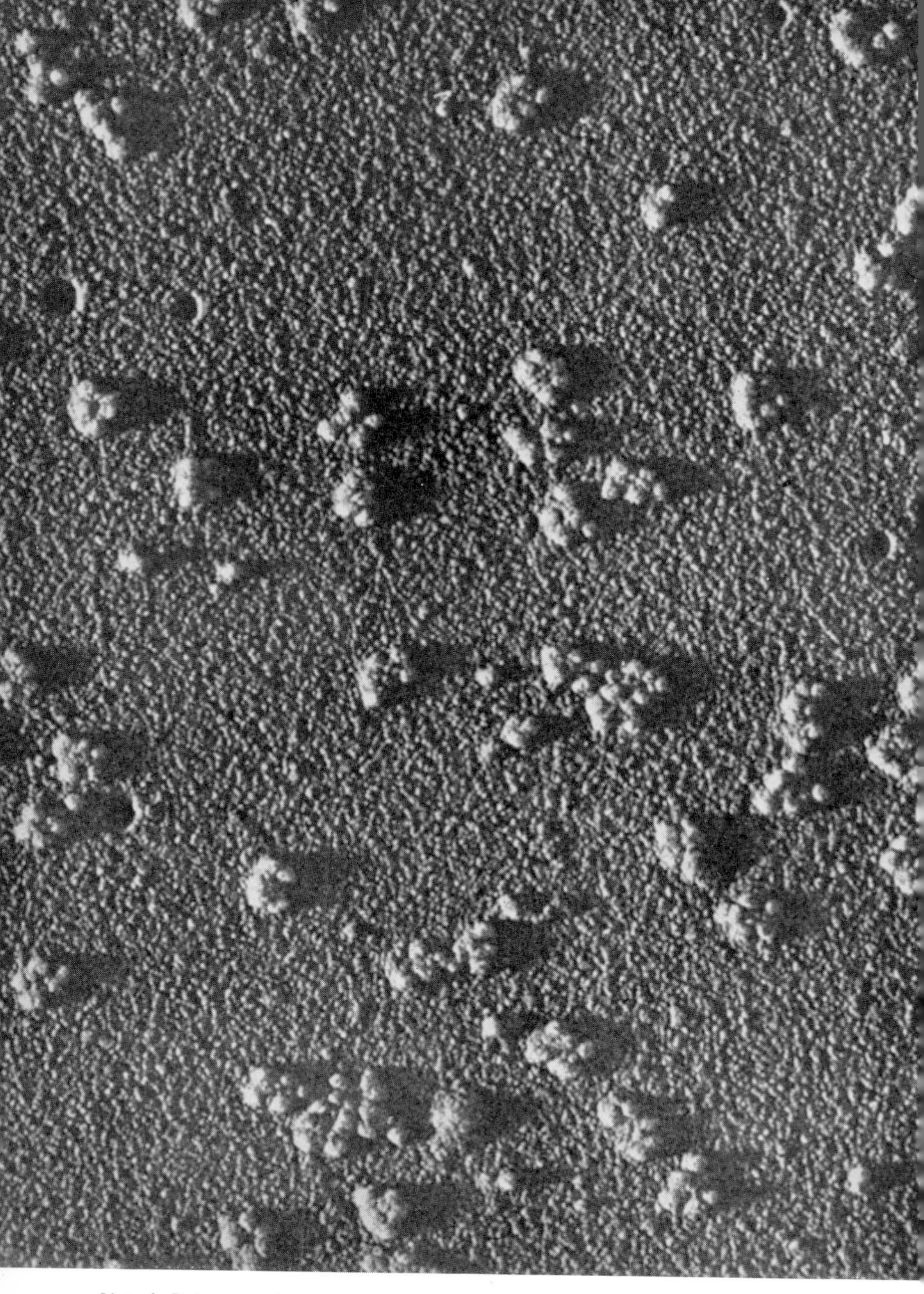

Plate 6. Polysomes from rabbit reticulocytes, shadowed with platinum (approx. $\times$ 200,000). (*Courtesy of Professor A. Rich*)

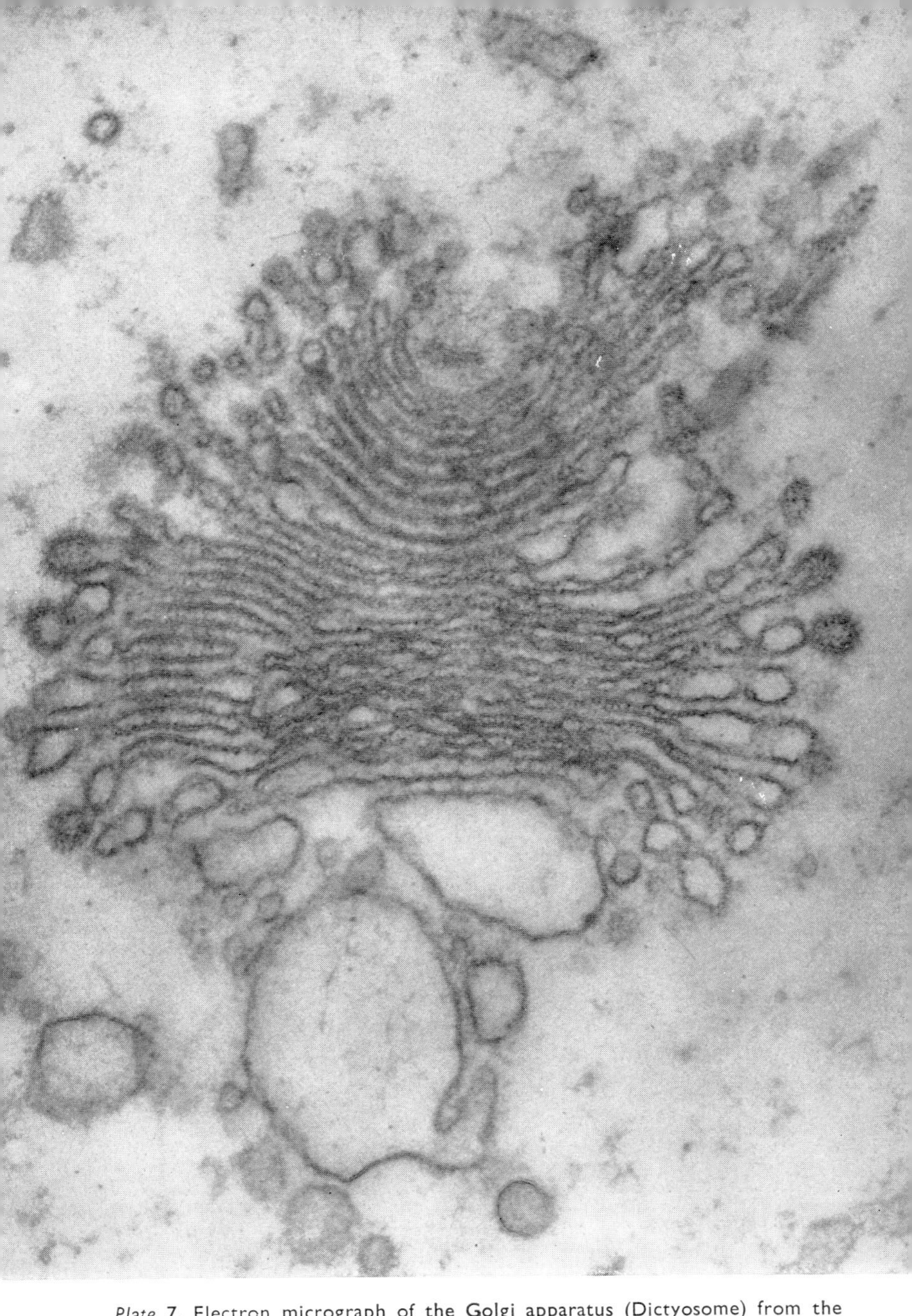

Plate 7. Electron micrograph of the Golgi apparatus (Dictyosome) from the protozoon *Mixotricha*. (*Courtesy of Dr A. V. Grimstone*)

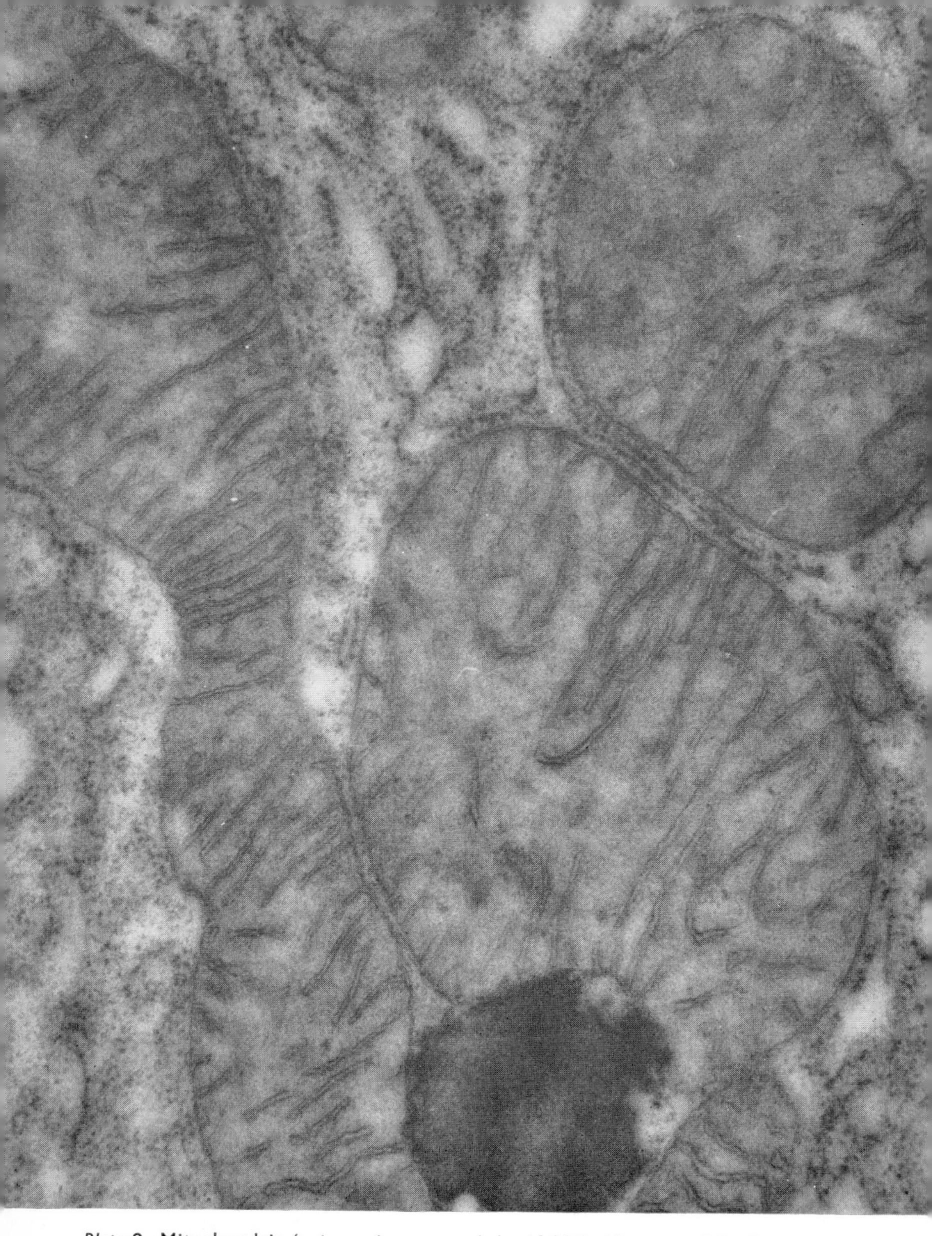

Plate 8. Mitochondria (guinea-pig pancreas) ($\times$ 46,000). (*Courtesy of Dr G. Palade*)

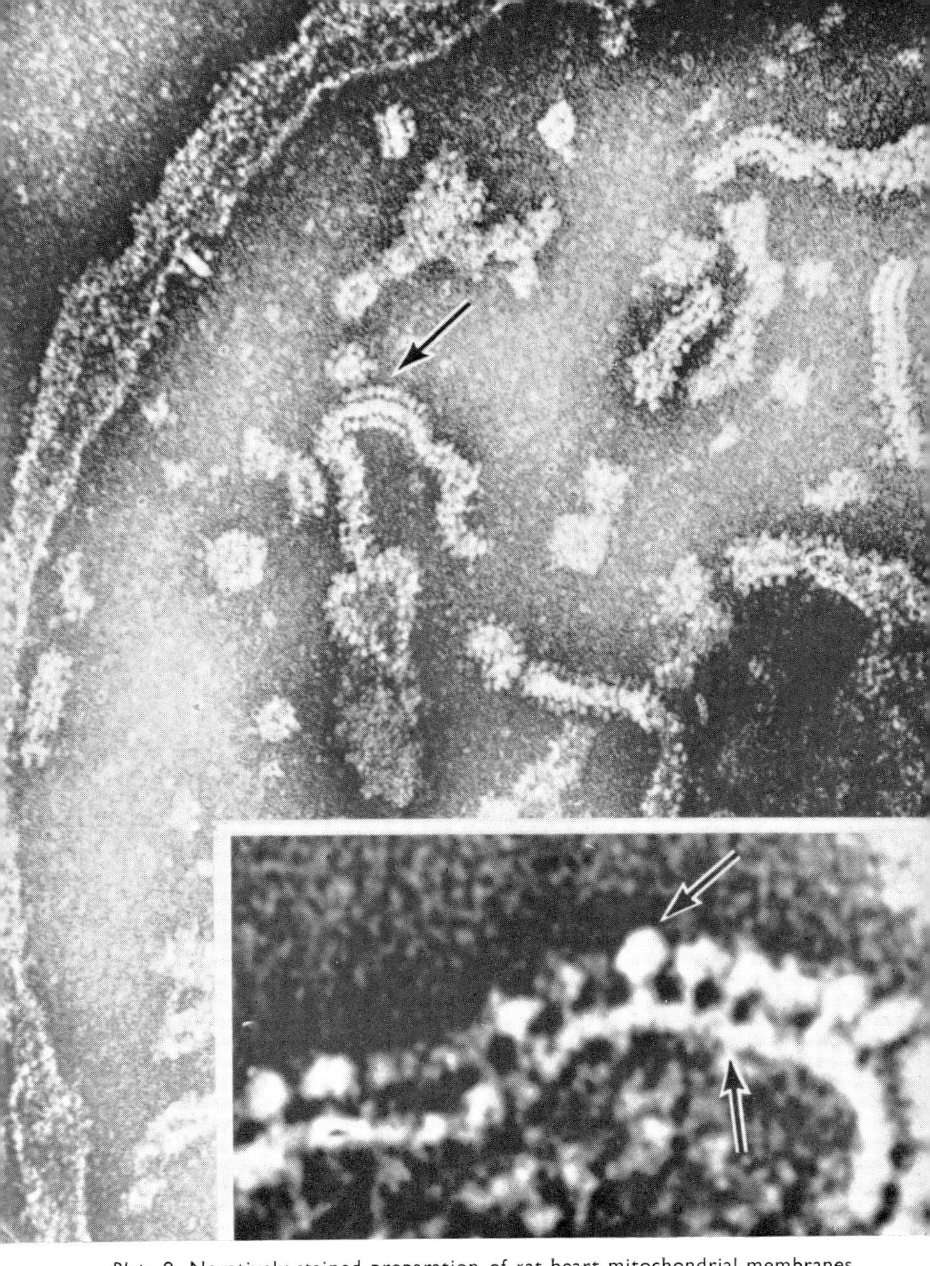

Plate 9. Negatively stained preparation of rat heart mitochondrial membranes (× 120,000) with (inset) an enlarged segment of a crista, showing elementary particles (× 600,000). (*Courtesy of Professor H. Fernandez-Moran*)

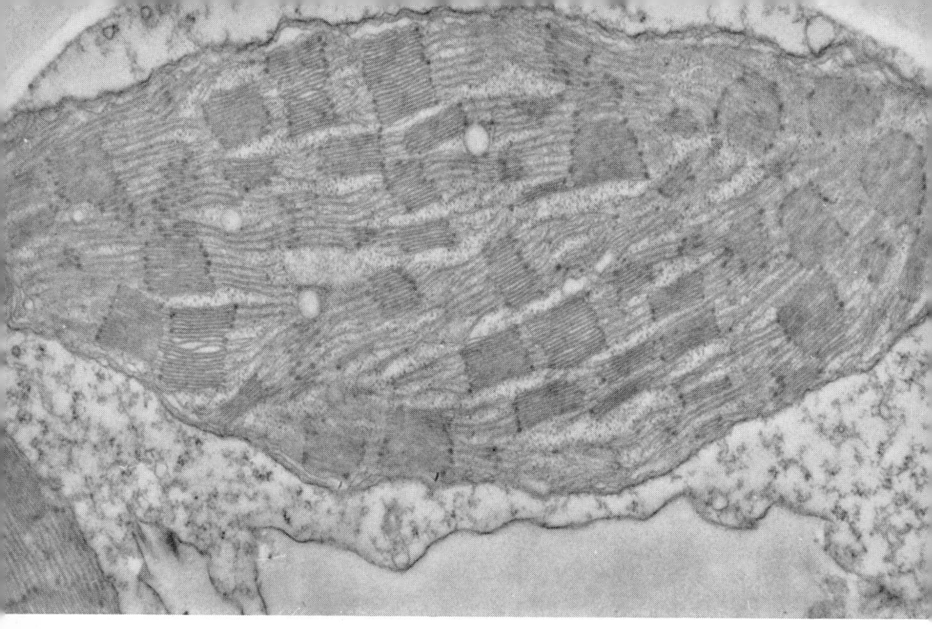

Plate 10A. Electron micrograph of a chloroplast showing its general structure (approx. $\times$ 30,000). (*Courtesy of Dr A. E. Vatter*)

Plate 10B. Electron micrograph of grana (approx. $\times$ 200,000). (*Courtesy of Dr A. E. Vatter*)

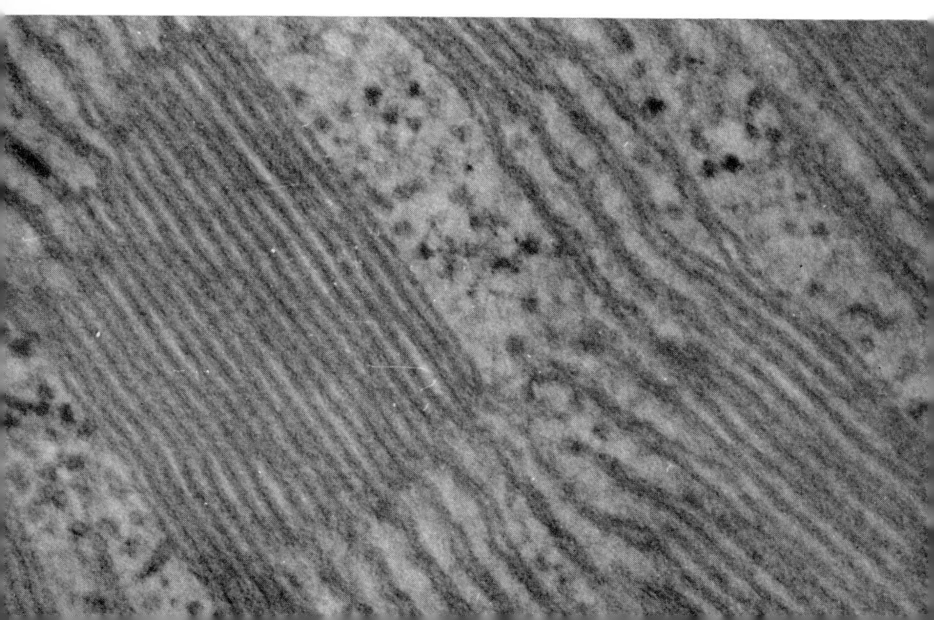

Plate 11A. Isolated lamellae from a disrupted chloroplast, negatively stained (×66,000). (*Courtesy of Dr R. P. Park*)

Plate 11B. Detail of the surface of an isolated lamellae, showing arrays of quantasomes (×330,000). (*Courtesy of Dr R. P. Park*)

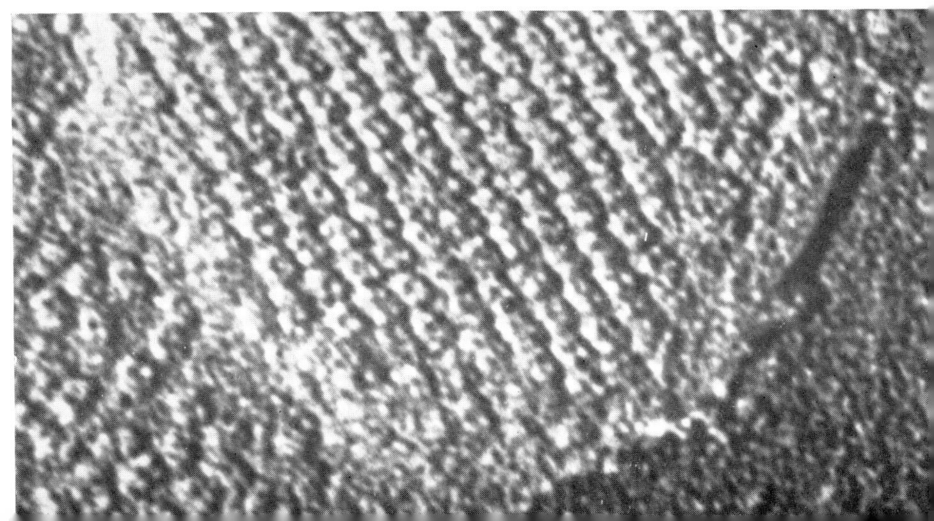

Plate 12. An electron micrograph of a cross-section of a myofibril. (See figure 5.7A.) *(Courtesy of Dr. H. E. Huxley)*

Plate 13A. An electron micrograph of a longitudinal section of striated muscle, showing the different banded areas (× 16,000). *(Courtesy of Dr H. E. Huxley)*

Plate 13B. A higher power photograph of a single sarcomere showing the thick and thin filaments and their relationship to each other and the *Z* band (× 46,000). (See figure 5.7B.) (*Courtesy of Dr H. E. Huxley*)

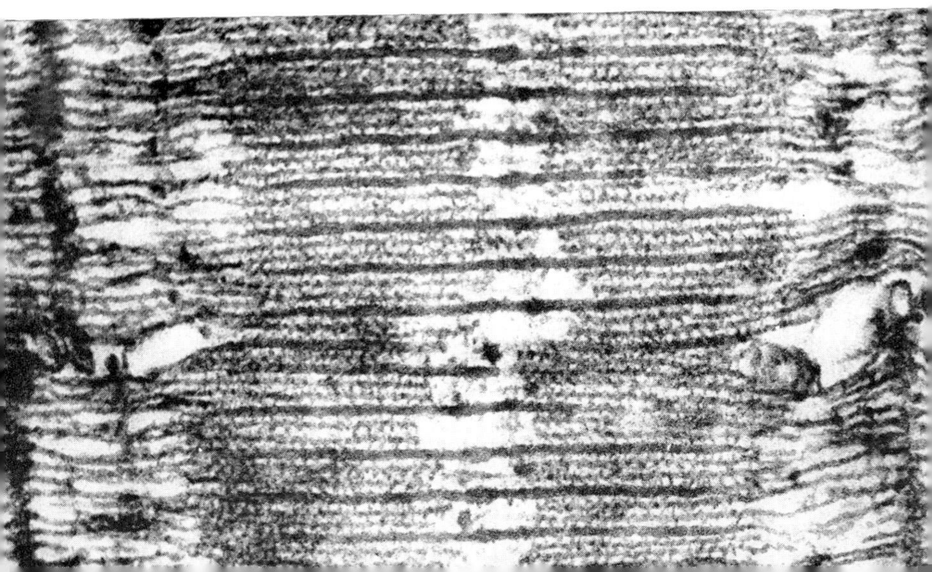

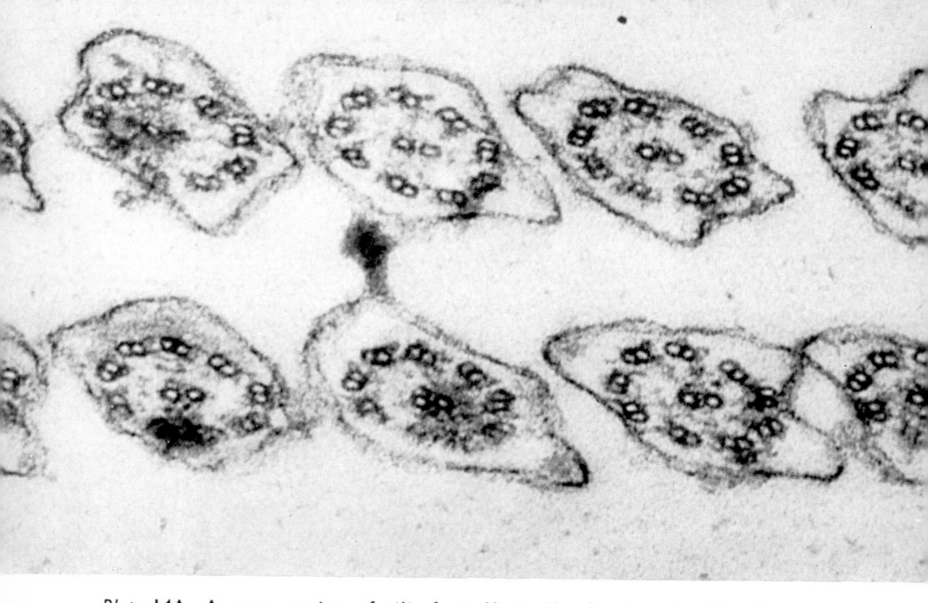

Plate 14A. A cross-section of cilia from *Vorticella*, showing the '9 + 2' structure (approx. × 120,000). (*Courtesy of Sir John T. Randall*)

Plate 14B. Centrioles in a cell of a human lymphosarcoma (× 85,000). (*Courtesy of Dr W. Bernhard*)

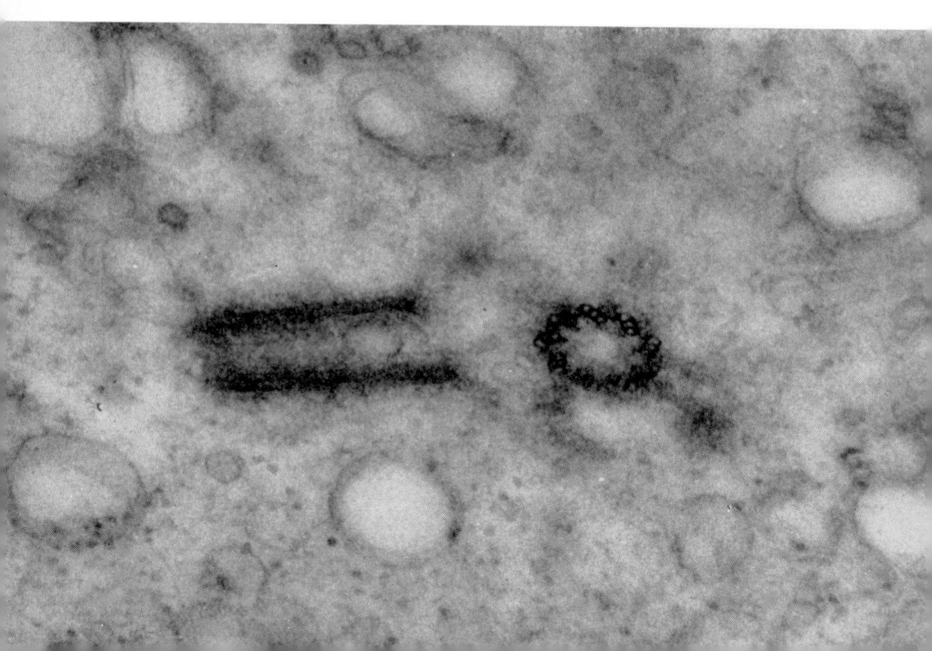

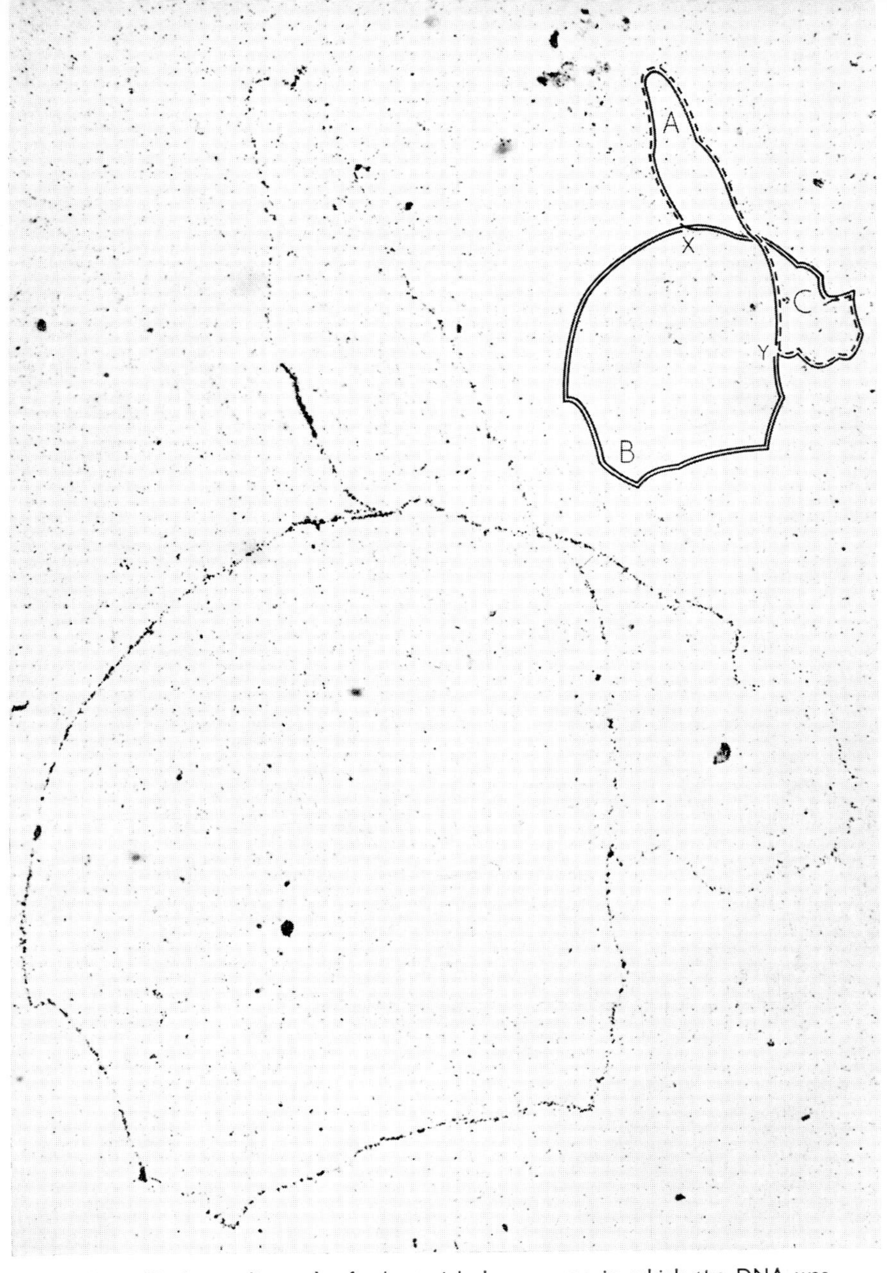

Plate 15. Autoradiograph of a bacterial chromosome in which the DNA was labelled with tritium. The inset shows diagrammatically how the structure consists of three sections (A, B, and C) that arise at the two forks (X and Y). The main loop is about 250μ in diameter. (*Courtesy of Dr J. Cairns*)

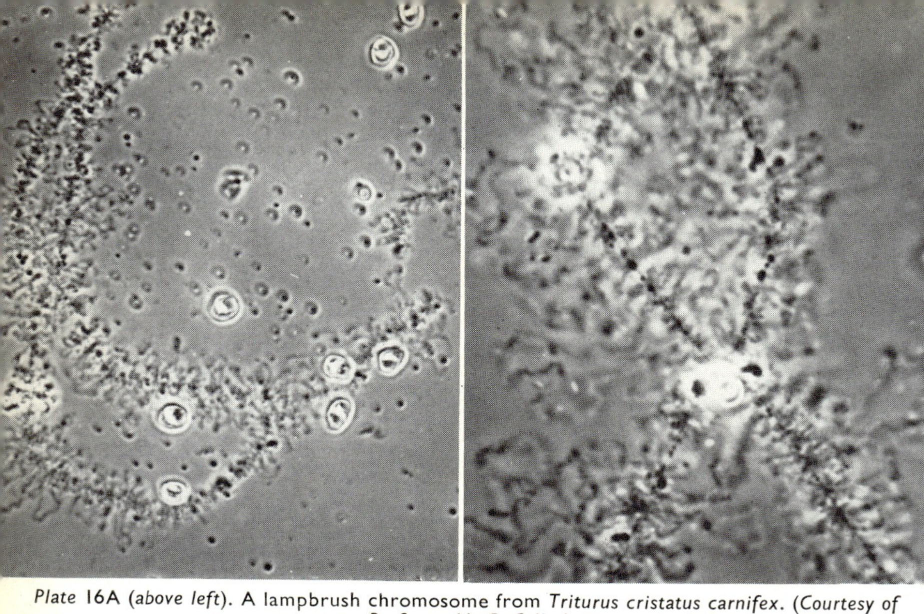

Plate 16A (*above left*). A lampbrush chromosome from *Triturus cristatus carnifex*. (*Courtesy of Professor H. G. Callan*)

Plate 16B (*above right*). Part of a lampbrush chromosome from *Triturus cristatus karelinii*, at higher magnification. (*Courtesy of Professor H. G. Callan*)

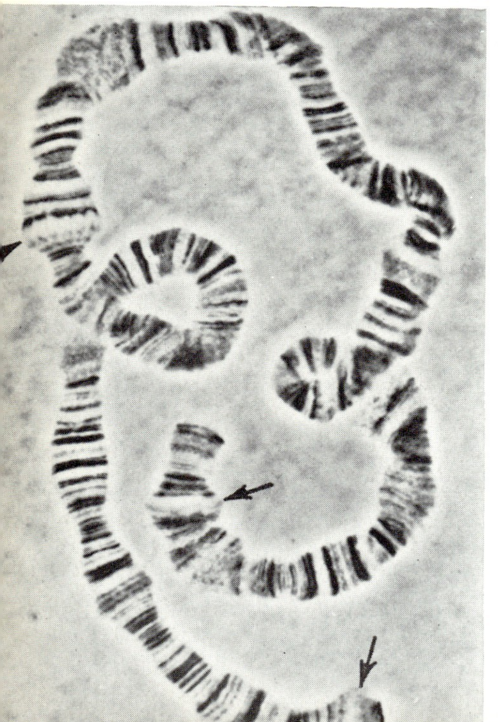

Plate 16C (*left*). A giant chromosome from *Smittia spec* showing prominent banding and several partly developed 'puffs' and (*below*) Plate 16D, a giant chromosome from *Smittia spec* showing a fully-developed puff (Balbiani ring) (× 1,250). (*Both by courtesy of Professor H. Bauer*)

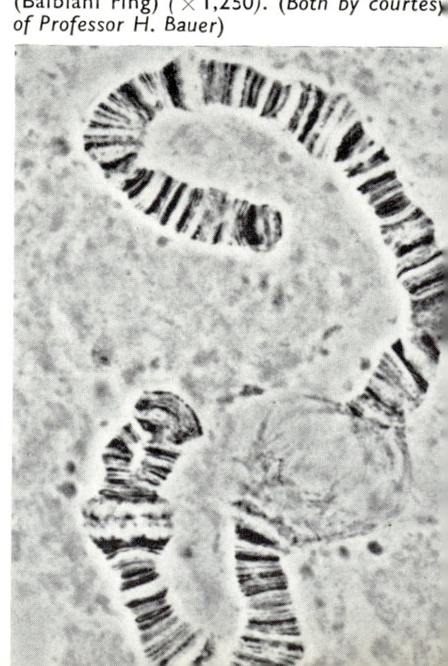

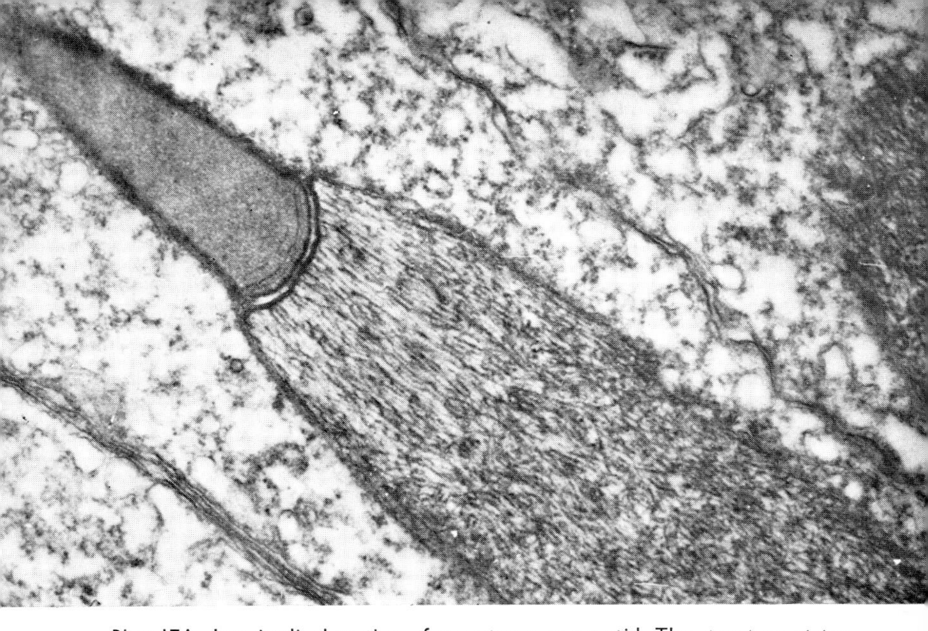

Plate 17A. Longitudinal section of an octopus spermatid. The structure at top left is the acrosome. The material stretching diagonally from it consists of orientated chromosomes. (*Courtesy of Dr H. Ris*)

Plate 17B. A higher magnification showing 100 Å fibres which can be seen to split into two 40 Å fibres in the encircled area. (*Courtesy of Dr H. Ris*)

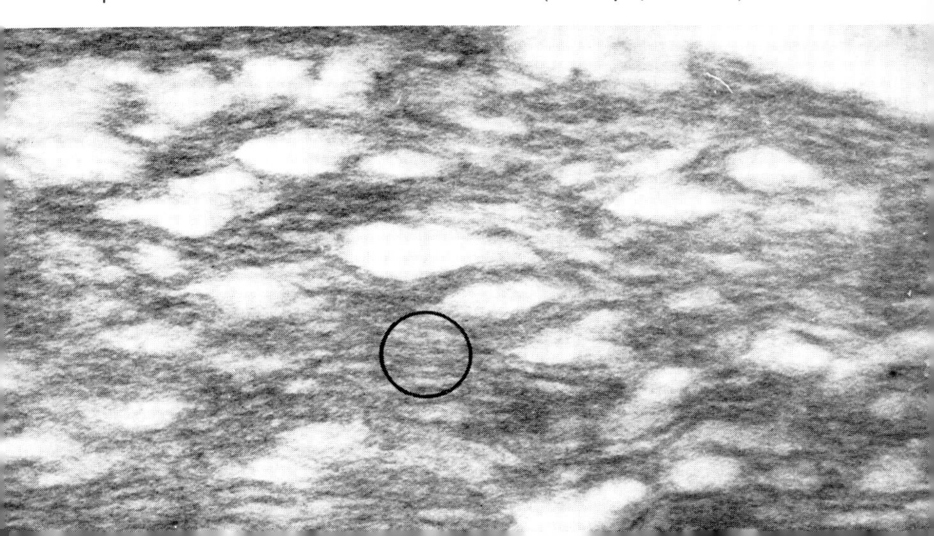

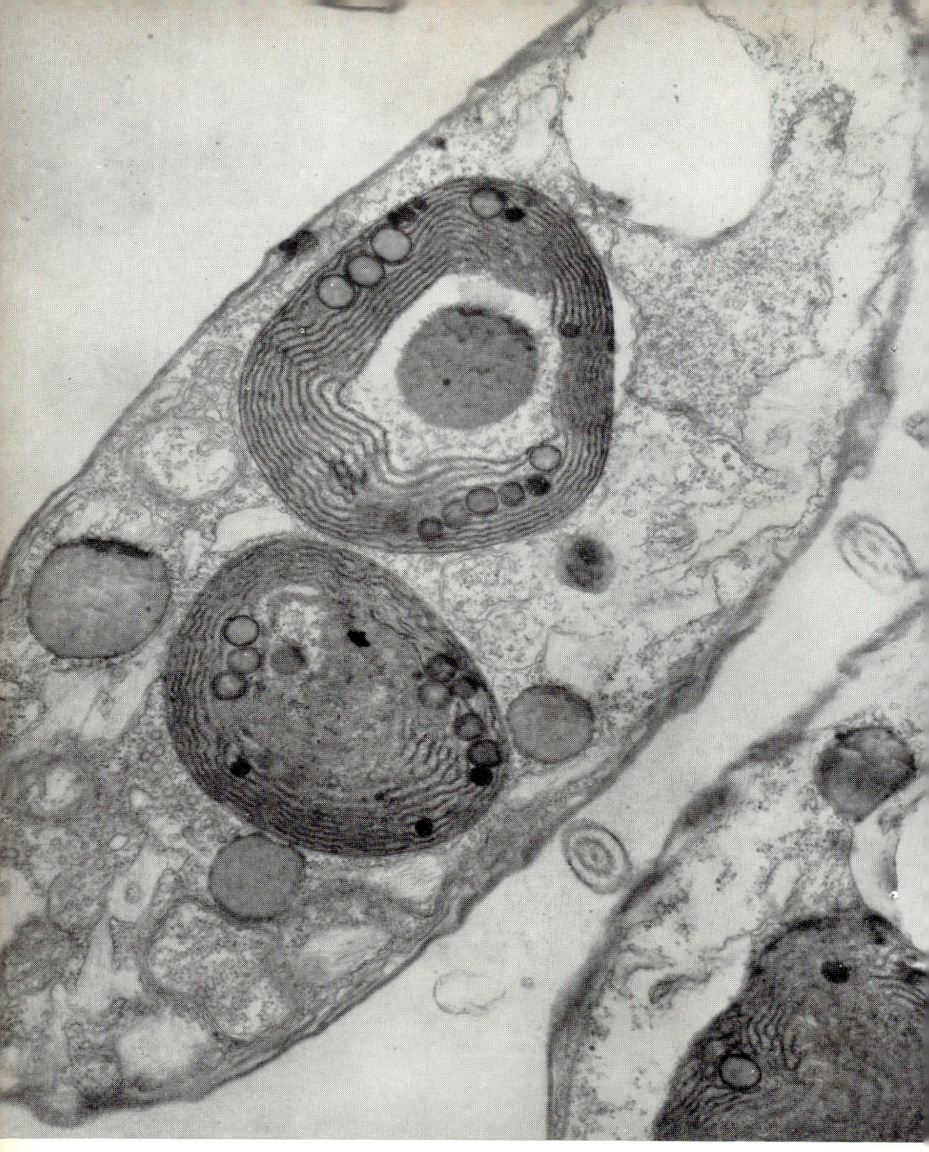

Plate 18. Electron micrograph of the single-celled organism *Cyanophora paradoxa*, containing blue-green algae which live symbiotically in its cytoplasm and hence function as chloroplasts. (*Courtesy of Dr W. T. Hall*)

phates are converted to triphosphates by a series of kinases. The enzyme DNA polymerase (in the presence of single stranded DNA as a primer) causes these triphosphates to link together with the loss of one molecule of pyrophosphate for each diester linkage. The

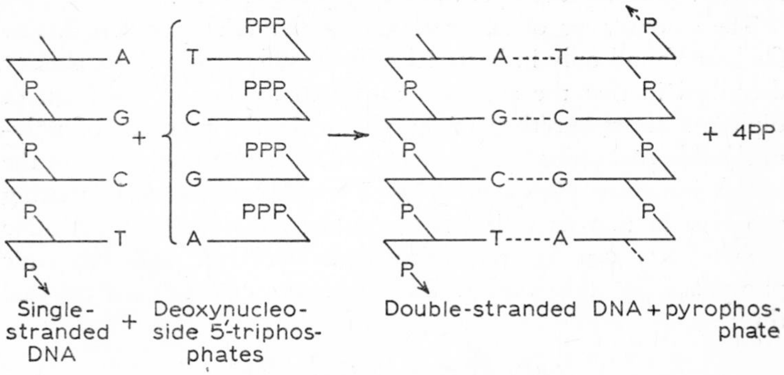

Single-stranded DNA + Deoxynucleoside 5'-triphosphates → Double-stranded DNA + pyrophosphate

Figure 6.3. **The synthesis of DNA by DNA polymerase in the presence of single-stranded DNA and a mixture of all four deoxynucleoside triphosphates.**

process is summarized in figure 6.3. The special steps involved in the synthesis of DNA are therefore relatively straightforward and (as was mentioned in chapter 2) it is thought that the nucleotides are lined up in order by the pairing of complementary bases with bases of the single-stranded DNA itself which thus serves as a primer for the reaction.

Ribonucleic acid. The synthesis of RNA is more complicated; there are at least three different mechanisms. One of these is strictly analogous to that which has been described for the synthesis of DNA. Ribonucleoside monophosphates are converted to triphosphates and these are caused to polymerize in the presence of RNA polymerase with the loss of pyrophosphate [20, 395]. The enzymes which perform this reaction are very widespread and occur in both higher and lower organisms. As in the DNA system, a primer, in the form of nucleic acid, is needed for polymerization to proceed. Ribonucleic acid is effective. However, it is of particular interest that similar enzymes have been found which can be primed by DNA [159, 491, 505]. It is possible that the two kinds of enzyme are the same and that the differences in priming capacity result from minor differences in the conditions used for measuring enzyme activity. This point has yet to be finally determined.

The second type of ribonucleotide polymerizing system has so far been found only in bacterial cells. It differs from those already described in that the activated nucleotides occur in the form of diphosphates rather than triphosphates. In the presence of polynucleotide phosphorylase the nucleotides are polymerized with the loss of inorganic phosphate [187]. This enzyme has an interesting property in that its action is reversible; consequently nucleotide diphosphates can be reconstituted from RNA and inorganic phosphate by this means. The reaction can be summarized thus:

$$n\mathrm{BRPP} \longleftrightarrow (\mathrm{BRP})_n + n\mathrm{Pi}$$

where B stands for a base, R for ribose and P for phosphate.

An important result is that some of the bond energy may be retained within the system, instead of being lost when RNA is broken down. Some people consider that the principal action of polynucleotide phosphorylase in the cell may be as an energy-conserving ribonuclease.

The remaining mechanism for the incorporation of nucleotides into RNA is a rather specialized one which is involved in the synthesis of soluble (transfer) RNA. It will be remembered that this RNA is distinguished by having a terminal -pCpCpA grouping. There is a special phosphorylating system in cell extracts which is responsible for adding these three nucleotide residues terminally to other RNA molecules [6, 77, 207, 527]. The precursors of the terminal nucleotides are triphosphates and pyrophosphate is released in the process.

One of the main problems in relation to RNA synthesis concerns the nature of the primer (that is, the template) in the living cell. In bacterial cells the situation is fairly straightforward since only three main types of ribonucleic acid have been recognized in them; ribosomal RNA (rRNA), messenger RNA (mRNA) and transfer

RNA (tRNA). Ribosomal RNA forms a very large part of the total RNA of most cells. It is one of the very stable components of cells and is probably almost as stable as DNA. The nature of the primary template for rRNA has been solved by the demonstration by Yankofsky and Spiegelman that there is a portion of DNA which is complementary to rRNA and which presumably carries the template for it [519, 520, 521]. Whether all the rRNA in a given bacterial cell is formed on this template or whether secondary templates are formed from which the rRNA in turn is synthesized is not yet known.

With regard to the origin of mRNA a good deal more is known. Messenger RNA is so-called because it is thought to be formed on DNA and is considered to be the template on which amino acids are lined up to form proteins [236]. The evidence for this is rather indirect but depends on the following observations:

(i) A hybrid DNA–RNA complex can be isolated from many kinds of cells. The RNA in this complex is very rapidly labelled when labelled precursors are added to the medium in which the cells are growing [58, 185, 419, 525].

(ii) Rapidly labelled RNA has similar base ratios to DNA of the same cell [161, 162, 485, 486] and specifically anneals with it [203].

(iii) This labelled material very rapidly appears again in association with ribosomes [58, 185].

(iv) When bacteria are inoculated with bacteriophage (which contains DNA) the synthesis of all bacterial enzymes stops and the energies of the infected bacterial cell are given over entirely to the synthesis of bacteriophage protein. At this time a rapidly labelled RNA appears in the cytoplasm which has base ratios exactly the same as the bacteriophage DNA [485, 486].

(v) RNA can be formed in the presence of a DNA primer in vitro and the base ratios of this RNA are identical with those of the DNA used as primer [159, 491].

This evidence all points to the formation of mRNA on a template of DNA. The question does not rest there, however, because it still remains to be seen how this can be accomplished. The obvious answer might be that mRNA is formed on single strands of DNA by base pairing, in exactly the same way as DNA is probably duplicated. However, this explanation encounters the difficulty that mRNA synthesis probably goes on almost continuously, whereas DNA is almost certainly in double-stranded (duplex) form most of the time. Consequently, other kinds of pairing have been suggested and systems have been proposed which would permit RNA to be formed around the DNA double spiral. Alternatively, it has been shown that bases can swing out from the inside of the DNA molecule and may thus become available on the outside. Finally, there is the possibility that sections of the DNA molecule may become unwound (like an untwisted piece of string) thus revealing single-stranded sections without the whole molecule having to be unravelled. There is now good evidence that only one of the two strands of DNA leads to the synthesis of mRNA in the cell [89, 474] and this is clearly relevant to the nature of the mechanism.

Most details of the synthesis of tRNA have now been worked out. As was indicated above, the terminal pCpCpA sequence is probably added in the cytoplasm. The difficulty which arises in postulating templates for tRNA is the existence of so many unusual bases in it. Some of these bases are, however, modified after the RNA chain has been formed [148]. It has been shown that sections of DNA exist in bacterial cells which are complementary to tRNA molecules [172]. Therefore, most of each tRNA molecule is probably formed directly on DNA by the action of RNA polymerase; the terminal pCpCpA sequence is added by special enzymes later and finally other special enzymes are responsible for the modification of many of the bases.

The nucleolus

In the more highly developed cells containing a nucleus the picture is more complicated because, in the nucleus alone, there are several kinds of RNA. Most of the complications are concerned with the nucleolus. This structure is very rich in RNA. Although all its functions have not been clarified it is well established that it is the site of ribosomal RNA synthesis [39, 68, 125, 378, 379, 380, 381, 382, 429]. It was recognized many years ago that nucleoli were always associated with specific sites on chromosomes, called the nucleolar organizers (297). More recently it has been shown that anucleolate mutants of *Xenopus laevis* lack the ability to synthesize rRNA (68). Moreover, DNA extracted from isolated nucleoli specifically hybridizes with rRNA. The evidence is, therefore, very strong that rRNA is made in the nucleolus; however, it may also perform other functions, such as the assembly of ribosomes or the modification of tRNA. It contains at least two kinds of RNA, distinguishable by having different base ratios and different turnover rates. One of these, with a very high turnover rate, has base ratios similar to the DNA of the same cell [161, 162, 426]. When cells are exposed to a radioactive precursor this RNA becomes labelled immediately after the RNA which is associated with DNA [426]. The other has base ratios similar to rRNA.

While some RNA is certainly synthesized in chromosomes [376], there is no conclusive evidence that all cytoplasmic RNA is synthesized in the nucleus. Some experiments point strongly in this direction [170, 171], but there is also evidence that much of the rapidly synthesized (possibly messenger) RNA never leaves the nucleus. Hence, the picture is more obscure than it is in bacterial cells.

There are still many problems to be cleared up and the beginner may find the position confusing. It can be summarized briefly as follows. It now seems likely that in normal cells the primary template for all RNA is DNA. Messenger RNA is probably copied directly from DNA. Ribosomal RNA is probably also copied directly but, on the other hand, it may be copied from a copy by some special mechanism. Part of the molecule of tRNA is probably formed on a DNA template and the primary molecule is then considerably modified.

Protein synthesis

Much of what has been said about nucleic acid synthesis has a direct bearing on protein synthesis [9, 199]. The association between the two was suspected for many years before it was positively demonstrated. In particular, Caspersson and Brachet emphasized that there was a close correlation between the RNA content of a tissue and its capacity for protein synthesis. The crucial role of the nucleic acids, particularly in ordering the sequence of amino acids in proteins, is now well established but, before considering this aspect of protein synthesis, the straightforward mechanism of polymerization must be discussed.

Amino acids are first activated by reacting with ATP in the presence of an activating enzyme (figure 6.4). However, whereas

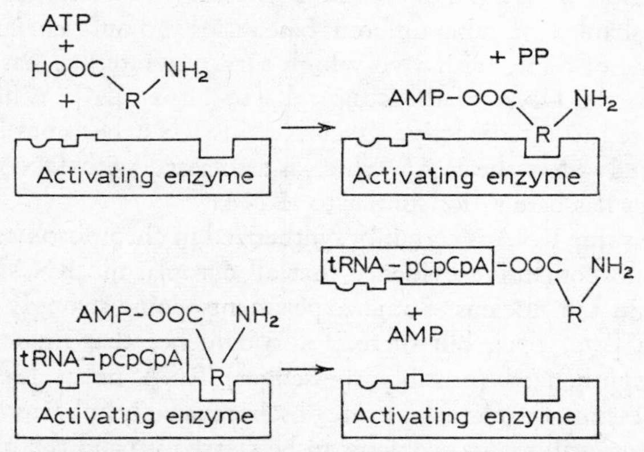

Figure 6.4. The role of activating enzyme in linking an amino-acid to a specific sRNA molecule. In the presence of ATP an amino-acyl adenylate is formed, which remains bound to the activating enzyme and reacts with sRNA at its surface. The enzyme is responsible for 'recognizing' the amino acid and the sRNA molecule and bringing them together.

in most reactions the terminal phosphate from ATP is added to the compound to be activated, this reaction is different in that the AMP residue is combined with the amino acid and pyrophosphate is released. The *amino-acyl adenylates* formed in this way are extremely reactive and, if mixed together in solution, will spontaneously form random polypeptides. This random condensation does not occur in living systems because each amino-acyl adenylate remains bound to the activating enzyme which catalyses its formation (figure 6.4). The activating enzyme then performs a second role by bringing a molecule of tRNA into contact with the amino-acyl adenylate on its surface. The tRNA and amino-acyl adenylate react, with the result that AMP is released and a compound is formed which consists of a tRNA molecule with a molecule of amino acid attached to the pCpCpA terminal sequence.

Each of the twenty amino acids has specific activating enzymes

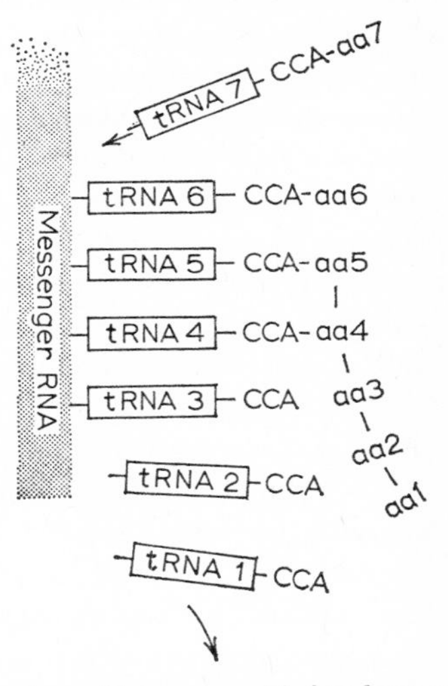

Figure 6.5. How amino acids are assembled to form a polypeptide. tRNA molecules act as adaptors and are assembled in order by pairing in a complementary fashion with bases on messenger RNA. After the peptide bonds are formed the tRNA molecules are released.

and specific tRNA. It must be clear that no specificity can reside in the terminal pCpCpA grouping common to all tRNA molecules. Consequently the activating enzyme itself must be responsible for bringing together an amino acid with its specific tRNA. It is postulated that the enzyme has two stereospecific sites, one for the amino acid and one for the tRNA. Strong evidence has been obtained for this contention by altering the structure of an amino acid after it has become bound to tRNA and subsequently showing that the altered amino acid is inserted into proteins in place of the correct one [79].

A complete series of tRNA's with attached amino acids is formed in the manner described. The next step in protein synthesis is the alignment of these on a template to form a protein molecule (figure 6.5). As has been mentioned, the template is mRNA. The amino-acyl tRNA can be bound by ribosomes, which also have the capacity to bind mRNA (256). It is well established that ribosomes play a crucial part in aligning amino-acyl tRNA molecules on an mRNA template and in the formation of polypeptide molecules. The probable mechanism is illustrated in figure 6.6. The growing peptide chain is attached to the tRNA molecule specific for the most recently added amino acid. This is located by the appropriate nucleotide sequence (represented by +) on the mRNA chain, and fills one of two possible 'slots' in the ribosome for tRNA. The other, vacant, 'slot' is over another nucleotide sequence (represented by O) which codes for the next amino acid. At 1 the appropriate amino-acyl tRNA is seen entering the 'slot'. When it has become located on the template the peptide chain swings from the peptidyl-tRNA to the new amino-acyl tRNA, as shown at 2. This leaves an unloaded tRNA molecule which is rejected (3). Finally, the ribosome moves relative to the mRNA until the peptidyl-tRNA is again in the first 'slot' leaving the other vacant (4). The whole cycle is then repeated and each time an amino acid is added to the peptide chain. Movement of the ribosome requires energy which is probably supplied by GTP (264). This process of peptide synthesis on a mRNA tem-

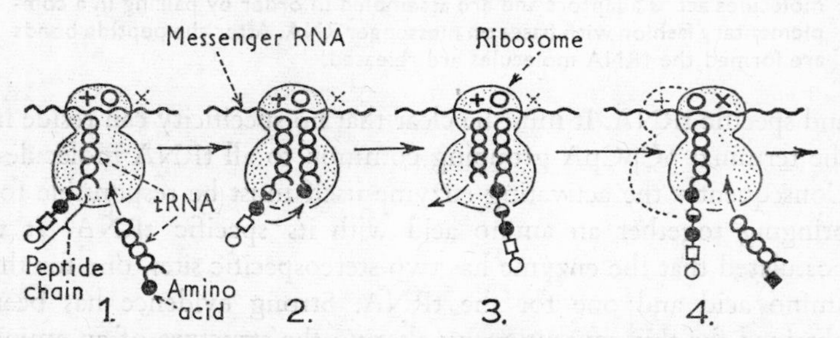

Figure 6.6. **The role of the ribosome in synthesizing peptides from amino acyl-tRNAs on a mRNA template. See text for details.**

plate is referred to as 'translation'. The mRNA molecule, like all nucleic acid molecules, is orientated, with a free 5′ group at one end and a 3′ group at the other. Translation starts at the 5′ end and proceeds to the 3′ end [188, 343, 416, 470, 471].

Each protein has, of course, an amino group at one end of the molecule and a carboxyl group at the other. These are respectively referred to as the *N*-terminal and *C*-terminal ends. It has been shown that amino acids are assembled on the template commencing at the *N*-terminal end [40, 41, 497]. As the chain is formed the *N*-terminal end may become detached, but the incomplete molecule always remains attached to the template by the most recently attached amino acid until it has been completed. Detachment of the complete protein requires ATP.

It follows from the above that 5′ ⟶ 3′ orientation of mRNA corresponds to N ⟶ C orientation of the proteins formed. In *E. coli* the first amino acid of each peptide is usually *N*-formyl methionine, for which there is a special tRNA (82). This is removed when the peptide has been made. In eukaryotes this special mechanism for chain initiation may be absent.

Each strand of mRNA becomes associated with about five or more ribosomes (forming a polysome, polyribosome or ergosome) during protein synthesis [352, 490] (plate 6).

All cells contain ribosomes and ribosomes from different sources are remarkably similar in structure. They have a molecular weight of three to four million and a sedimentation constant of 70–80 S (S = Svedberg unit, a measure of the sedimentation velocity of particles in the ultracentrifuge) They are about 200 Å in length and 100–150 Å wide. They are composed of RNA and protein, of which about 60 per cent is RNA. In low magnesium concentrations the bacterial ribosome can be made to break into two unequal portions of 30 S and 50 S respectively. The 30 S particle but not the 50 S particle alone can bind mRNA. Bacterial ribosomes are smaller than ribosomes from animal cells (MW 2·5 million as compared with about 5 million). The subribosomal particles of animal cells are correspondingly larger, about 40 S and 60 S, and whole animal ribosomes are about 80 S.

Polysomes may be found free in the cytoplasm, and this is the rule in bacteria. In higher cells too they are commonly found in this form but in them they are also typically found associated with a reticulum together with which they form the endoplasmic reticulum or ergastoplasm [359, 360, 361, 390, 392]. This consists of a lipoprotein double membrane with ribosomes dispersed along it. In most cells the endoplasmic reticulum is represented by a few vacuoles but in cells which secrete large amounts of protein, such as the pancreatic acinar cell the reticulum is highly developed and may fill almost the entire cell with tightly packed membranes (see plate 5). The 60 S components of the ribosomes associate with the membranes in such a way that polypeptide chains are formed within the cisternae while the 40 S components and mRNA remain on the other side.

The relationship of the endoplasmic reticulum to other membranes in the cell has aroused a good deal of speculation, particularly since it is frequently continuous with the outer layer of the nuclear membrane. In certain cells it is continuous with the cell membrane itself and it has been proposed that channels pass from the cell membrane to the nuclear membrane through the reticulum. Since not all cells have a continuous reticulum, however, it seems unlikely that this is a general state of affairs.

The endoplasmic reticulum is often referred to as rough membrane by contrast with smooth membranes which do not possess ribosomes. Smooth membranes are often highly developed in cells which do not form proteins in large amounts, but which have other secretory functions, for example, the steroid secreting cells of the adrenal cortex and sex glands [133]. In nearly all cells some smooth membranes can be demonstrated in the juxtanuclear zone and these have been identified with the Golgi apparatus (plate 7) [37, 175, 176, 388, 438]. This apparatus is also particularly prominent in protein-secreting cells and very often protein granules can be seen within it, for example, zymogen granules in pancreatic cells. The Golgi apparatus is now thought to be a collecting zone for secretory granules; it may be continuous with the endoplasmic reticulum. In the pancreas, there is excellent evidence that protein secreted into the cisternae of the endoplasmic reticulum later accumulates in the Golgi apparatus where it gives rise to the zymogen granules. Portions of the Golgi apparatus containing the granules bud off and migrate to the cell membrane where the granules are excreted by fusion of the enveloping membrane with the cell membrane, followed by evagination [78] (chapter 9).

The pathway of protein synthesis which has been described is the one which has been most thoroughly studied and in which most of the steps have been well authenticated. However there is some evidence for an alternative pathway involving the so-called 'S-protein', but not activating enzymes. Although little is known about this system it is not too unlikely that alternative modes of protein synthesis exist since certain inhibitors of protein synthesis affect the formation of some proteins but not others [192].

Part Four: The Organization of Cellular Activity

7. The Control and Integration of Function

It was observed many years ago that the many metabolic pathways in cells must be carefully regulated, since very few intermediary metabolites are formed in excess. For instance, in bacterial cells the number of intermediary metabolites may run into hundreds, and yet only a few substances are found to accumulate in the medium in which they are grown. Furthermore, the composition of individuals in a rapidly growing population of cells remains virtually constant. Yet, if the rate of formation of one factor were only fractionally greater than that of others, it would eventually become enormously larger than them. These simple observations bespeak special mechanisms for maintaining the harmonious functioning of all parts of a cell in order that no one property can outrun the others. These regulatory mechanisms [7], often referred to as homeostatic mechanisms, have been found to operate in much the same way as control processes in engineering practice, where they are called servo-mechanisms.

In chapter 1 it was indicated that the simple properties of enzymes themselves contributed to adjustment to environmental variations. The mechanisms will now be discussed in greater detail.

Enzymically catalysed reactions

The cell is in a state of dynamic equilibrium and all its components (except possibly DNA) are continuously being built up and broken down. Consequently the composition of any cell is a function of the relative rates of all the reactions going on within it.

The rate of any reaction is in general dependent on the frequency of collisions between reacting molecules and therefore on the concentrations of the reactants and products. In the presence of a catalyst (usually an enzyme in a biological system) the collision rate is effectively increased. A few biological reactions can occur spontaneously in the absence of enzymes but the vast majority are enzymically catalysed and, other things being equal, the reaction rates are dependent on the amount of enzyme present [E]. Besides being proportional to the amount of enzyme, the velocity [V] of an enzyme reaction is also a function of the concentration of the substrate [S], inhibitors [I] of various kinds, and physical factors such as pH and temperature. The relationship can be represented as follows

$$V = k[E] . f[S][I] \text{ (pH, temperature etc.)}$$

The three factors which are important in the control of metabolism are the concentrations of enzyme, substrate and inhibitors. In chapter 1 it was shown that, if the amount of an enzyme is constant, then variations in substrate concentration lead to variations in enzyme activity. Over a wide range of concentrations the velocity of a reaction is proportional to substrate concentration but, eventually, a maximum is reached beyond which increasing the substrate concentration has no effect. If the normal substrate concentration is such that it lies on the steeply sloping part of the Michaelis–Menten curve (figure 1.2) then a simple measure of automatic control already exists since increases or decreases of substrate concentration lead to corresponding changes in the rate of substrate utilization.

Similar to this simple process of substrate control is the phenomenon of product inhibition. Many reactions are progressively inhibited by the accumulation of end products and, where the reaction is reversible, there is an inverse relationship between the concentration of the product and the velocity of the reaction. This constitutes a simple regulating mechanism since accumulation of products leads to a slowing down of the reaction and consequently decreased production of products.

Feedback control

These simple types of rate regulation undoubtedly play an important part in the control of cell metabolism but probably of greater importance are the more complicated mechanisms. The simplest is directly related to product inhibition and is called feedback inhibition [164, 174]. However, whereas in product inhibition the inhibitor is the immediate product of the reaction, in feedback inhibition it is a distant one. Many examples have been discovered in both bacterial [229, 309, 478, 495, 516, 523, 544] and animal cells [42, 61, 62, 205, 245]. For example, as shown in figure 7.1, several of the enzymes involved in the synthesis of uridine monophosphate are inhibited by products of later reactions in the sequence.

In the mechanisms so far considered the amount of enzyme need

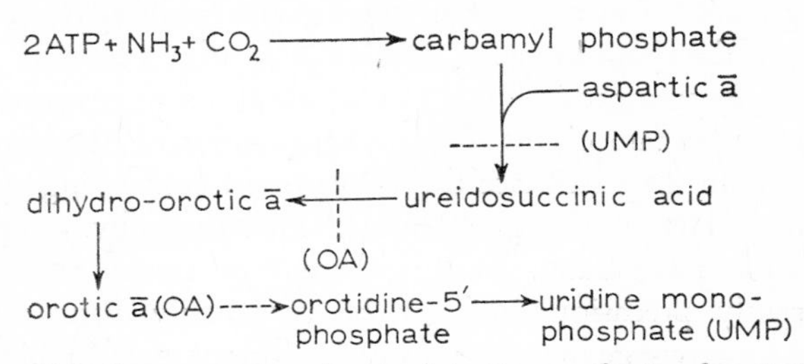

Figure **7.1. Some reactions involved in the synthesis of pyrimidines, showing points at which enzymes are inhibited by distant products.**

not vary. However, in two very important systems of metabolic control the amount of enzyme itself varies. In the first, *enzyme induction*, the enzyme is caused to increase when its substrate accumulates. In the converse process, *enzyme repression*, the amount of enzyme diminishes when its products accumulate. If the product is a distant one the process is called *feedback repression*. The mechanisms of enzyme induction and repression will be discussed in detail later in this chapter.

Limiting reactions. Consider a chain of reactions, represented as follows, in which the capital letters represent metabolic pools and the small letters represent enzymes

$$A \xrightarrow{a} B \xrightarrow{b} C \xrightarrow{c} D \ldots N$$

All the enzymes in this series may operate at substrate concentrations well below their maxima, in which event the rate of the entire series fluctuates according to the substrate concentration presented to the first reaction. On the other hand one of the reactions may go more slowly than the one preceding it, in which event the substrate of the slow reaction will tend to accumulate. Consequently this reaction will very rapidly come to operate at its maximum velocity. The result will be that, although subsequent reactions are potentially capable of operating faster, this reaction will limit the rate at which substrate becomes available to them and hence constitute a limiting reaction. The kinetics of all the reactions following it may then be described in terms of this one enzyme. Limiting reactions usually occur either at the beginning of a whole chain or, if the chain branches, immediately after a branch.

There are three possible ways in which limiting enzymes could operate to maintain chains of enzyme reactions in harmony with each other. In the first place the velocities of controlling enzymes might be predetermined genetically so that they always operated at the same fixed rates. Such a system of initial rate control is inherently precarious and can be very easily disturbed. It is difficult to envisage its playing an important part in biological adaptations.

The second possibility is that the relative rates of two or more limiting reactions might be kept in balance by competition for a common substrate or co-factor. This kind of control (substrate limitation) does play a part in biological processes and one or two examples will be discussed. The third possibility is that the rates of limiting reactions might be controlled by their products, in other words by feedback, either feedback inhibition or repression.

Substrate limitation. When the supply of substrate to an enzyme system is restricted subsequent reactions are slowed automatically. This happens in the reactions following a limiting reaction. Also, in certain circumstances, a substance may be used preferentially by one pathway. An alternative pathway may be deprived of it and hence be controlled simply by limitation of substrate when demands on the main pathway are heavy. Figure 7.2 shows how protein synthesis, pyrimidine synthesis and the urea cycle compete for common pools. Depletion of the pools by a high rate of either protein or pyrimidine synthesis tends to diminish the excretion of ammonia by the urea cycle. The functional value of this kind of control is obvious. If the urea cycle did not automatically adjust in this way then amino acid deficiencies would become apparent at much higher levels of nutrition.

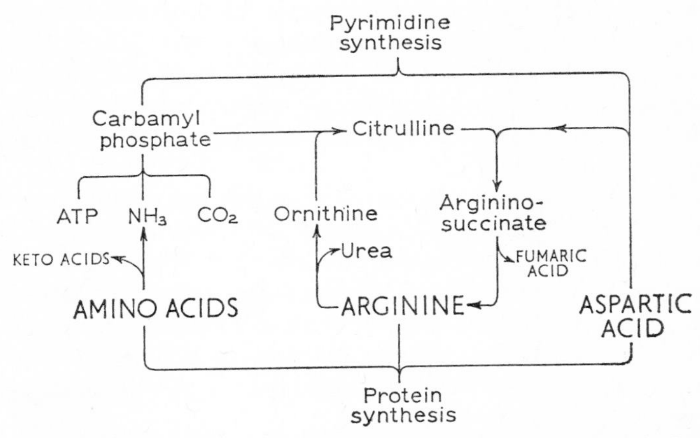

Figure **7.2. Some reactions competing for common substrate pools. Protein synthesis may divert substrates from the urea cycle or compete with synthesis of pyrimidines.**

When enzymes have similar affinities for the same substrate, competition for the common substrate or co-factor may provide the basis of a regulating mechanism. The example which has attracted most interest and speculation is the phenomenon known as the Pasteur effect. This is the increase in glycolysis which occurs when facultatively aerobic organisms are moved from aerobic conditions to anaerobic conditions. It is observed in bacteria and also in animal cells. It can be explained by competition for ADP or inorganic phosphate by the glycolytic pathways on the one hand and the Krebs cycle (or corresponding aerobic pathway) on the other (figures 4.3, 4.4). When the aerobic pathway is operating it competes very successfully for available ADP and inorganic phosphate, either of which may be limiting. As a direct consequence the amount of ADP or inorganic phosphate available for ADP-requiring reactions in glycolysis is limited. Consequently, glycolysis is diminished.

A similar effect, the Crabtree effect (sometimes referred to as the inverse Pasteur effect) can probably be explained by a similar mechanism. In this, cellular respiration is inhibited by increasing the amount of glucose in the medium. It is assumed that the increased glucose concentration leads to increased glycolysis and increased utilization of the available ADP or inorganic phosphate. Alternative explanations for both these effects have been advanced but they all depend on the same principle of competition for a common substrate or co-factor.

Certain other substrates or co-factors in cells may be limiting. In particular NAD may not be present in sufficient quantities to saturate all the enzyme systems for which it is required. Consequently, an increase in some NAD-requiring reactions may result in an inhibition of others. Where two or more metabolic pathways depend on a common metabolite this element of competition may enter, for example, between glycolysis and the pentose shunt and between RNA and DNA synthesis.

Feedback inhibition

Simple feedback inhibition is probably one of the commonest means of controlling metabolic pathways. The role of feedback inhibition in pyrimidine synthesis has been indicated (figure 7.1) and similar mechanisms are widespread. A common result of feedback in one pathway is diversion of metabolites to another. For example in the pyrimidine synthetic pathway feedback inhibition may lead to accumulation of carbamyl phosphate which may then be diverted into the urea cycle (figures 7.1, 7.2). Another example of interactions between linked pathways, illustrated in figure 7.3, shows the feedback relationship existing between the interconversion of some of the purine nucleotides and the synthesis of histidine.

Some of the most important relationships of this kind are connected with nucleic acid synthesis [111, 397, 398]. These are summarized in figure 7.4.

A particularly interesting example of feedback inhibition, which gives some indication of the refinements of the control mechanisms involves aspartokinase in *Escherichia coli* (figure 7.5) [448]. This enzyme catalyses the formation of aspartylphosphate from aspartic acid and ATP. Aspartylphosphate then undergoes a series of reactions to give rise to lysine, threonine and methionine. In certain microorganisms there are, in fact, two different aspartokinases, one

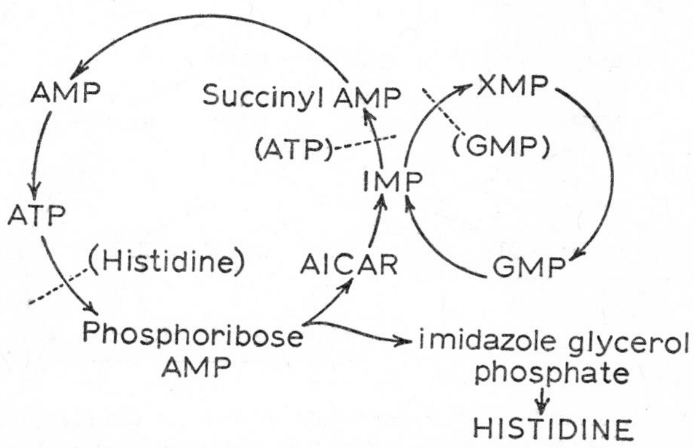

Figure 7.3. Feedback relationships in the utilization of a common substrate (IMP) by different pathways. Accumulation of one group of products in excess tends to divert the substrate into other pathways. Operation of competing pathways simultaneously results in a balance between them. (IMP – inosine monophosphate; AMP-adenosine monophosphate; ATP – adenosine triphosphate; AICAR – 4-amino-5 imidazole-carboxamide riboside; XMP – xanthosine monophosphate; GMP – guanosine monophosphate.)

of which is inhibited by lysine and the other by threonine. If either of these amino acids accumulates in excess, the supply of aspartylphosphate is reduced but not stopped. Other feedback controls occur later in the sequence of reactions to limit the formation of each individual amino acid. In this way, not only is the excessive formation of each amino acid completely controlled, but the supply of a distant precursor is adjusted simultaneously.

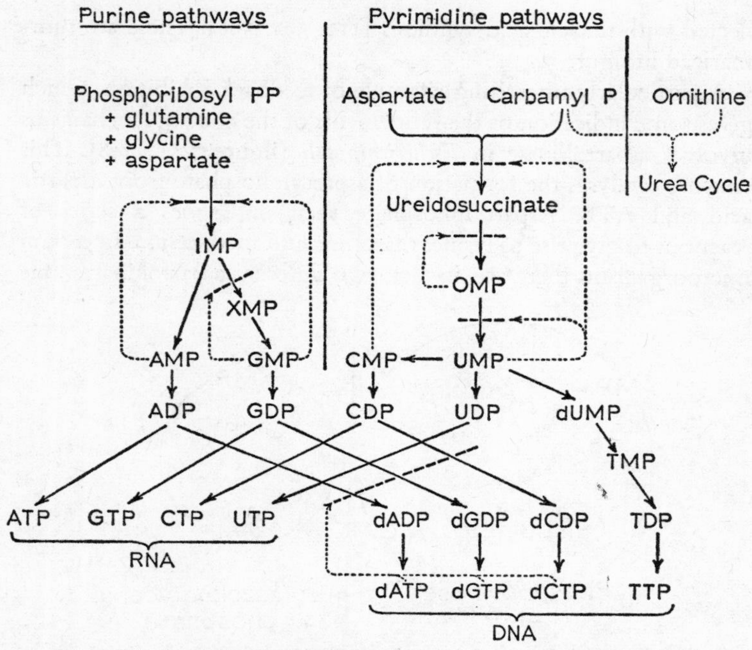

Figure **7.4.** Some feedback relationships involved in the regulation of RNA and DNA synthesis. (I – inosine; X – xanthosine; A – adenosine; G – guanosine; O – orotidine; U – uridine; C – cytidine; T – thymidine.)

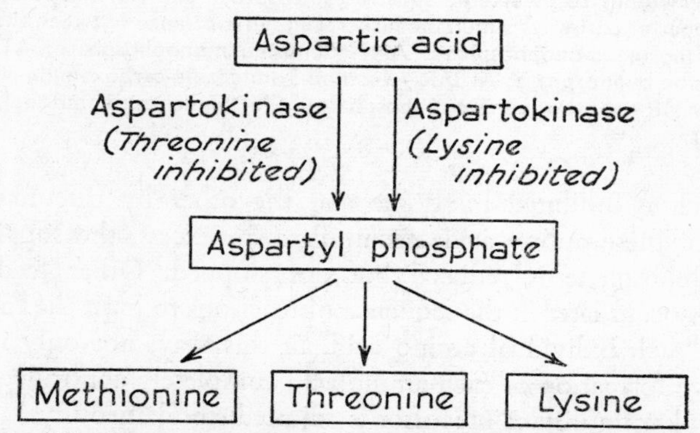

***Figure* 7.5. The way in which the synthesis of aspartyl phosphate is regulated by different end-products through two isozymes of aspartokinase.**

Enzyme repression and induction

These processes [246, 303, 394, 484] involve changes in the amounts of enzymes and their effects are somewhat different. Feedback inhibition operates instantaneously. Repression on the other hand, while it achieves essentially the same end, takes some time to become effective. In bacteria the delay may be relatively short. For instance, in *Escherichia coli* the addition of tryptophan to the medium may result in an almost immediate cessation of synthesis of tryptophan synthetase. Similarly, an accumulation of inorganic phosphate may rapidly lead to cessation of synthesis of alkaline phosphatase [158]. In animal cells, however, the repression of D-glutamyltransferase by glutamine may take some hours to become effective [365]. Furthermore although the formation of the enzyme rapidly stops in these cases, enzyme already present in the cell may continue to act for some time before it is degraded or ultimately diluted out by cell growth. Thus enzyme repression is a much more sluggish process than feedback inhibition.

Many of the enzymes involved in feedback inhibition are also controlled by repression. For instance, some of the enzymes of pyrimidine synthesis, which are subject to feedback inhibition, are also repressed by the same products (figure 7.1) [524].

In enzyme induction enzyme synthesis is stimulated by increasing the substrate concentration. The phenomenon has many similarities to repression and it is generally considered that the two processes have the same basis. Repression may be the commoner phenomenon but induction has been studied in much more detail and these studies form the basis of our present understanding of the two processes. The most celebrated example of enzyme induction involves β-galactosidase in *Escherichia coli*. Jacob and Monod and their colleagues [246, 324, 325] have shown that when phenyl-β-galactoside or a similar substance is added to a medium in which wild-type *E. coli* are growing then the enzyme β-galactosidase is immediately produced in large amounts. When the substrate is later removed, enzyme production ceases very quickly (but not immediately as was at one time thought [43]). From a study of the kinetics of induction of this enzyme, and of a variety of mutants,

Jacob and Monod have proposed a system of regulation of enzyme synthesis which has received a great deal of support from other recent work [246].

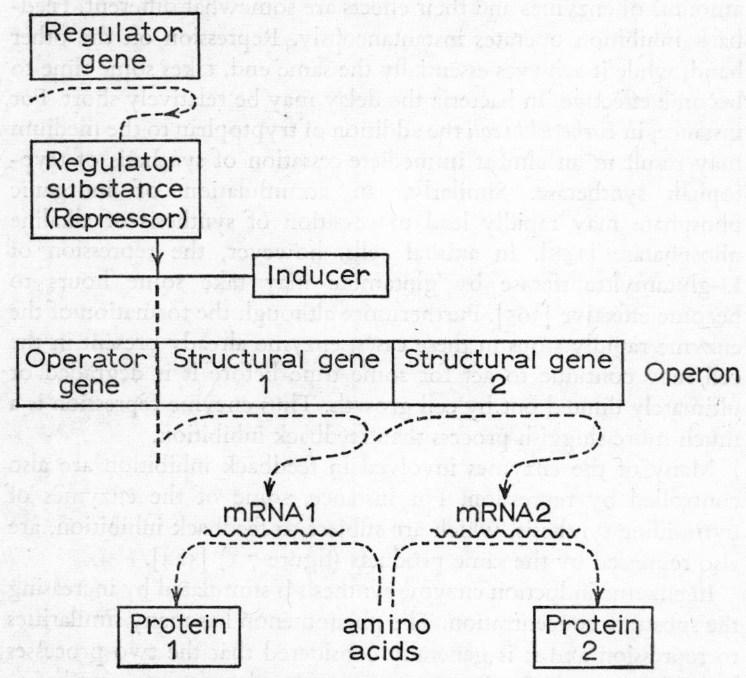

Figure 7.6. The scheme proposed by Jacob and Monod for regulation of protein synthesis in microorganisms. Structural genes give rise to messenger RNA which triggers the synthesis of specific proteins. The formation of messenger RNA by structural genes is initiated by an operator gene, which can be blocked by a repressor, formed by yet another gene, a regulator gene. The repressor substance can be inactivated by an inducer and protein synthesis then proceeds. In its absence the repressor prevents protein synthesis. A complex of an operator gene and the structural genes it controls is called an operon.

The system which they propose is illustrated in figure 7.6. It is suggested that the rate of formation of a protein is directly related to the rate of formation of mRNA and that this is, therefore, the primary rate-controlling factor in protein synthesis. An uninhibited *structural gene* is thought to produce mRNA, and therefore enzyme, at a maximal rate (constitutive enzyme production). Since the enzymes β-galactosidase and galactoside transacetylase are simultaneously induced by the inducer for either of them, it is postulated that the structural genes for these two enzymes are themselves controlled by a master gene called the *operator gene*. A complex consisting of an operator gene and the structural genes controlled by it is called an '*operon*'. The parts of an operon occupy the same locus or immediately adjacent loci in the chromosome. In the β-galactosidase system there is another gene, the *regulator gene* which produces a substance called the *repressor* or *regulator* which acts directly on the operator gene to prevent it functioning. It is supposed that the inducing substrate reacts with the regulator to prevent it from repressing the operator gene. The consequence of adding an inducing substrate is therefore the production of an increased amount of enzyme.

The evidence advanced in support of this system is as follows. By application of orthodox segregation methods of genetic analysis (chapter 8), the Pasteur group has obtained direct evidence for the participation of three separate genes in the synthesis of β-galactosidase. One of these, the *y* gene has to do with the entry of substrate to the cell [84] and is not directly concerned with the regulatory mechanism. It will not be considered further. The other two genes are the *z* gene or structural gene for β-galactosidase, and the *i* gene or regulator gene. Mutations of the *z* gene result either in the production of no enzyme at all or in the production of an altered enzyme. Mutations of the regulator gene, on the other hand, lead to the production of a very large amount of enzyme, since the regulator substance is either not formed or is abnormal and inactive. Uncontrolled synthesis of enzyme resulting in this way from mutation of the regulator gene is called constitutive synthesis. By crossing and recombination it has been shown that the *i* and *z* genes are separated from each other by a considerable distance on the bacterial chromosome.

Evidence for the operon concept has also been obtained from genetic mapping techniques. These show that the structural genes for enzymes which can be induced simultaneously with the same inducer are very close to each other. The genes for β-galactosidase and galactoside-transacetylase appear to be adjacent to each other, and the operator is immediately adjacent to them. Several other examples of operons have been described. One of the most striking involves the eight or nine structural genes concerned with the enzymes responsible for histidine synthesis.

Originally it was thought that each molecule of mRNA was responsible for the synthesis of one molecule of protein, because the synthesis of β-galactosidase seemed to start and stop immediately the substrate was introduced or removed. It has now been shown that one molecule of messenger RNA is responsible for the synthesis of some ten to twenty molecules of enzyme [241, 293]; and more recent studies of the kinetics of induction show that there is a short lag period which is compatible with this figure (43). Regulation of protein synthesis by this type of mechanism has been investigated very intensively, and there is now very good evidence for it. However, different mechanisms may also operate and, for instance, there is evidence that regulation may occur at the ribosomal level [518].

These mechanisms have been studied most intensively in microorganisms, but clearly they are of very great interest in cells of higher organisms in relation to cytodifferentiation. Inducible and repressible enzymes have now been described in many animal tissues and cells. In particular, the evidence for the inducibility of tryptophan pyrollase is now very good [270] while tissue culture studies have provided evidence for the inducibility of the thymidine kinases [506], alkaline phosphatase [93, 94, 95] and arginase [269] in animal cells. Repression of glutamyltransferase in animal cells has been studied in some detail [116, 364, 365].

The enzyme tryptophan pyrrolase [180, 270] provides the most intensively studied example of adaptive behaviour in animals and well illustrates its complexity. Tryptophan pyrrolase activity in rat liver can be caused to increase within a few hours in response to two different kinds of stimuli. A rapid increase follows the administration of tryptophan; this is due to stabilization of the enzyme and is not prevented by blocking RNA synthesis. A slower increase follows the administration of cortisone; this is apparently due to increased mRNA synthesis as it can be prevented by blocking RNA synthesis with actinomycin D. After a few hours the level of tryptophan pyrrolase falls back to the normal level again. This appears to be due to active removal of the enzyme; which in turn depends on RNA synthesis since it can be prevented by blocking it after the enzyme is induced. Hence, at least three, and possibly more, mechanisms are involved in regulating the amount of active enzyme in this instance.

The response is generally much slower and smaller in animal cells than in bacteria. This is probably related to the greater stability of polysomes in animal cells. However, examples of immediate inhibition of protein synthesis following inhibition of synthesis of mRNA have been reported in animal cells [109] and hence this may not hold as a generalization.

Cyclic behaviour

Biological systems are characterized, not only by a delicate balance and integration of the many component reactions, but also by the occurrence of many examples of cyclic behaviour, for example the cell division cycle, the menstrual cycle, the periodic cycles of recurrent fevers like malaria, and so on. Some examples of these so-called Circadian rhythms are so precise that they are described as 'Biological Clocks'. In many which have been studied in detail it is found that there are regular increases and decreases in the activity of some enzymes. Hence, it is suspected that the same kinds of feedback controls may be involved.

The situation is in many respects analogous to that encountered in mechanical and electronic control systems in engineering. Most of these operate by feedback of information from the output side to the input side of a device in much the same way as a reaction product may regulate its formation by governing the rate of the reaction. It is a well-known feature of engineering control systems that if they are inadequately damped they tend to oscillate. This tendency is sometimes used, particularly in electronic circuits, to produce oscillations, which are cyclical events. A standard way to do this is to introduce a time-delay between the output and input side (for example, in electronics by a capacitor). This means that the damping signal from the output does not affect the input until it has overshot the equilibrium point. Consequently an oscillation results.

Biological oscillations or cycles are almost certainly produced in the same way, although reliable proof for this statement has yet to be obtained. Suggestive evidence is plentiful, however. For instance, the substance actinomycin D inhibits mRNA synthesis rather specifically and this has been found to inhibit one kind of biological clock [259], suggesting that it operates through an inducible or repressible enzyme system.

8. Reproduction and Heredity

When most growing cells reach a certain size they divide to form two daughters which are almost identical with the parent. To ensure this, all the parts of the cell have to be replicated and redistributed more or less equally. Furthermore, all parts have to be replicated to the same extent; otherwise the progeny would diverge progressively. Hence the whole process of replication is a carefully balanced complex of reactions which almost always conforms to a clearly defined pattern.

In many somatic animal cells the complete cycle from one cell division to another takes between twelve and twenty hours. Actual cell division (from the beginning of mitosis) occupies less than an hour. Protein and ribonucleic acid may be formed continuously between mitoses but DNA is usually synthesized for a limited period of about six hours and there then follows a gap of about four hours before cell division. The result is that the amounts of DNA, RNA and protein are exactly doubled during the single cycle. The stages of DNA synthesis are classified as the synthetic phase (*S*), the first gap (G1), between cell division and the onset of DNA synthesis, and the second gap (G2), from the end of *S* to the beginning of mitosis.

During cell division cell material is distributed to daughter cells in two ways. Most free molecules are probably distributed quite randomly; some of the cell organelles such as mitochondria and ribosomes may behave similarly. In some of the simplest organisms DNA may be redistributed in a more or less random fashion too, but in most lower organisms and all the higher organisms DNA is organized in the chromosomes and there are very precise mechanisms for their distribution to daughter cells. These mechanisms are, of course, mitosis and meiosis.

The structure of chromosomes

Despite extensive research on the structure of the chromosome [120, 128, 155, 261, 262, 320, 321, 335, 336, 405] much remains to be clarified. It is not difficult to fit most observed phenomena of gene replication into the system of DNA replication by base pairing. However, it is only in lower organisms that there is any evidence at all for the occurrence of free DNA molecules which could behave in this simple way [72, 530]. In such organisms, for instance bacteria, there is no well-defined nucleus and the DNA occurs as fibrils about 25 Å in diameter [263]. As discussed later, genetic evidence proves that each DNA molecule in lower organisms represents a series of genes arranged end to end. From the genetic point of view therefore it has a structure similar to a chromosome. Indeed, the term bacterial chromosome is frequently used to describe a group of linked genes in bacteria, although some authors prefer an alternative term, such as genophone, reserving the term chromosome for the structures found in the cells of higher organisms. The bacterial chromosome is in the form of a closed ring and there is evidence that replication begins at a specific starting point and progresses from there round the chromosome (plate 15) [71, 337, 526].

It is in relation to true chromosomes that it becomes clear that the simple concept of DNA replication is an oversimplification. The structure of chromosomes is indeed quite complex and has not yet been completely resolved. Furthermore, it is difficult to reconcile the apparent structure with genetic rules and even with mitosis.

Whereas the bacterial chromosome may consist only of DNA the chromosomes of animal and plant cells contain basic proteins (such as histones) in amounts equivalent to the DNA present. They also contain other, 'acidic', proteins in variable amounts, enzymes, such as DNA and RNA polymerases, and RNA.

During mitosis or meiosis chromosomes are much condensed and are readily visible by ordinary optical methods. However, during the greater part of the cell cycle they are greatly hydrated and are not visible in most cells, even when special staining methods are used. One or two exceptional instances are known, however, in which interphase chromosomes are visible, and these have been of great value in studying their behaviour and structure. In particular, the giant chromosomes of the salivary glands of certain diptera are easily seen, apparently because the fibrils of which they are composed have replicated time and time again while continuing to lie alongside each other. This is known as polyteny. As many as two to four thousand fibrils may occur in each giant chromosome, instead of the normal thirty to sixty. The result is a broad-banded structure which is readily visible and easily studied (plate 16C, D).

The other interphase chromosomes which have provided useful experimental observations are the so-called lampbrush chromosomes of amphibian oocytes (which probably occur in other species also). These chromosomes (plate 16A, B) are readily visible in interphase and can even be dissected out. They are seen as long strands with symmetrical side loops.

Chromosomes in general have some common features (figure 8.1). First, all normal chromosomes have a constriction which represents the attachment of the spindle fibres during cell division; this is known as the *centromere*. Chromosomes are classified according to the position of the centromere, for instance metacentric chromosomes have the centromere in the middle whereas telocentric chromosomes have it at one end. Secondly, most chromosomes divide length-wise and the two halves formed in this way are called *chromatids*. Thirdly, when chromosomes are stretched, as in lampbrush chromosomes, each chromatid can be seen to consist of one or more threads, called *chromonemata*, which have bead-like areas of increased density dispersed along them. These are known as *chromomeres*, and in the lampbrush chromosomes the side loops extend out from them. The bands in salivary gland giant chromosomes of diptera probably also correspond to chromomeres.

It has proved exceedingly difficult to investigate the ultrastructure of chromosomes at levels of high resolution because of their great complexity. This is largely the result of coiling and super-coiling of the elementary fibrils of which the chromosome is composed.

However, lampbrush chromosomes and also the chromosomes in the sperm cells of cephalopods (plate 17) are largely uncoiled. By using material like this, in conjunction with stereoscopic electron-microscopy, Ris [404, 405] has convincingly shown that the fundamental unit is a fibril 100 Å in diameter which is itself made up of two 40 Å fibrils which correspond to DNA-histone molecules. Ris calls the 100 Å fibril the *elementary chromosome fibril.* In most chromosomes the individual chromatid seems to consist of some sixteen to thirty-two of these fibrils lined up side by side.

Since the 40 Å fibrils are thought to represent nucleoprotein molecules [511] this structure presents a problem to the geneticist. It has been suggested that the fibrils represent different DNA-histone molecules folded alongside each other, but there is some experimental evidence to suggest that in fact they may represent a multi-stranded structure in which the subunits are identical [22, 100, 374, 476]. Experiments, in which the radioactive decay of chromosomes of amoebae which had incorporated P^{32} was studied, have been interpreted as indicating that the chromosomes contain at least sixteen subunits [154]. Autoradiographic studies on the replication of plant cell chromosomes suggest that these are also polynemic, i.e. made up of several identical strands [373]. This question remains to be clarified.

Understanding of the further organization of the chromosome is confused by the fact that the results of experimental investigations are conflicting. Many of these experiments have involved treatment of chromosomes with proteases and nucleases. It is generally agreed that ribonuclease causes some change in the chromosome but does not alter its general structure. Deoxyribonuclease, on the other hand, does not cause rupture of the giant salivary chromosome of diptera whereas it does cause fragmentation of the lampbrush chromosome [301]. The latter observation suggests that DNA runs longitudinally in the chromosome and is responsible for the continuity of its structure. It has been pointed out that in polytenic dipteran chromosomes there are so many strands lying alongside each other that, unless breaks occurred at exactly the same points in each fibril, the chromosome might not rupture and this could explain the discrepancy. There is other evidence, from tracer labelling studies, that the DNA molecule does in fact lie longitudinally in the salivary gland giant chromosome. The conclusion that DNA forms a continuous strand in the lampbrush chromosome is supported by the appearance on Feulgen staining and also by the electron microscopic studies of Ris. The weight of evidence thus supports this structure. However, each chromosome has many replication sites [386] and therefore some kind of 'linker' between individual replicative units (or replicons) may have to be postulated.

The most serious discrepancy concerns the action of proteolytic enzymes. Most authors maintain that these cause disruption of the chromosome [262], but in some very careful recent studies MacGregor and Callan [301] have challenged this conclusion.

The general structure which can be proposed provisionally from these studies is summarized in figure 8.1. The elementary chromosome fibril of 100 Å probably consists of two DNA-histone molecules arranged side by side and possibly cross-linked through histones or other proteins. These fibrils may run continuously the length of the chromatid or subunits may be joined end to end by non-histone protein. The fibrils may be replicated many times in each chromatid. The individual fibrils are probably randomly coiled, the whole mass of the chromatid coiled again and the whole structure coiled once more on top of this. In lampbrush chromosomes (plate 16A, B) it is thought that the DNA strand runs continuously along the chromonema, takes part in some complex coiling in the chromomere, then extends out into each side loop and returns to join the chromomere before continuing in the chromonema. The basic proteins, usually histones, seem to be associated with the nucleic acids throughout the entire length of the chromosome. There is good evidence, however, that RNA is concentrated at certain areas. In lampbrush chromosomes it is demonstrable in some of the side loops and in giant chromosomes (plate 16C, D) it is particularly associated with structures called *Balbiani rings* or *puffs* which appear to be expanded chromomeres. These are thought to represent functional areas of the chromosome. Chromosomes also contain large amounts of acidic proteins, sometimes called chromosomin. The spatial arrangement and functional significance of these

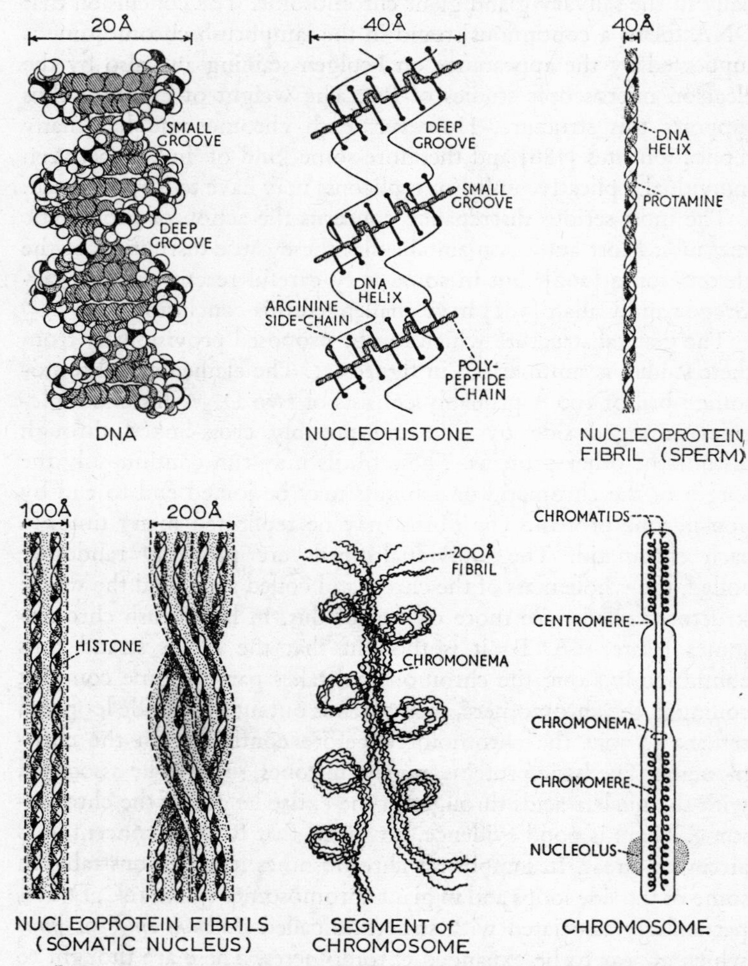

Figure **8.1. The way in which a chromosome may be constructed from nucleoproteins, based on electron microscopic and chemical evidence. (Modified from a diagram by Dr Hans Ris.)**

is at present completely unknown, although they may be present in amounts equivalent to histones.

With such a complicated structure it would be surprising if the chromosomes of higher cells behaved like the DNA molecules in Meselson and Stahl's experiment on *E. coli* (page 78). However, there is very good evidence that this is precisely what happens. Taylor and his colleagues [465, 466, 467, 468] found that when plant cells were exposed to radioactive thymidine all the chromosomes were labelled at first. At the next cell division (the tracer having been withdrawn) half the chromosomes were labelled and half unlabelled, as would be predicted if whole chromosomes behaved exactly like the DNA molecule in the Watson–Crick system of replication. No entirely satisfactory model has been proposed to reconcile all the findings.

Behaviour of chromosomes in cell division

In most multicellular and many single-celled creatures there is an alternation in the life-cycle between cells with two homologous sets of chromosomes (diploid) and cells with one set of chromosomes only (haploid). The haploid cells are usually the gametes or sex cells; they are formed from their diploid precursors by the process of *meiosis*. The diploid cells are the somatic cells; they are formed from diploid precursors by *mitotic division*. Mitosis can also occur in haploid cells and in certain forms, especially algae and mosses, a considerable and even major part of the life cycle may take place in the haploid state (the haplophase). In this case the somatic cells are haploid. Occasionally cells with more than two sets of chromosomes may arise giving triploid, tetraploid or polyploid forms. Whole organisms are occasionally tetraploid but polyploid cells are usually abnormal. Occasionally very abnormal forms, with random numbers of chromosomes, are found and these are classified as aneuploid.

Among the protista many organisms remain haploid throughout the entire life-cycle and only in special circumstances become diploid or polyploid.

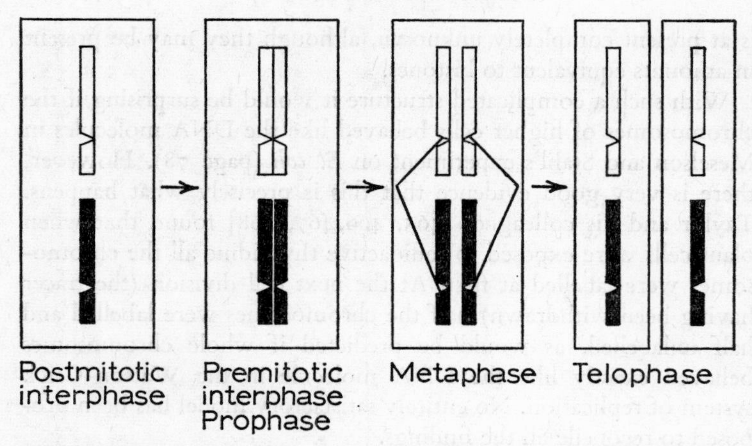

Figure 8.2. **The behaviour of a pair of chromosomes during mitosis.**

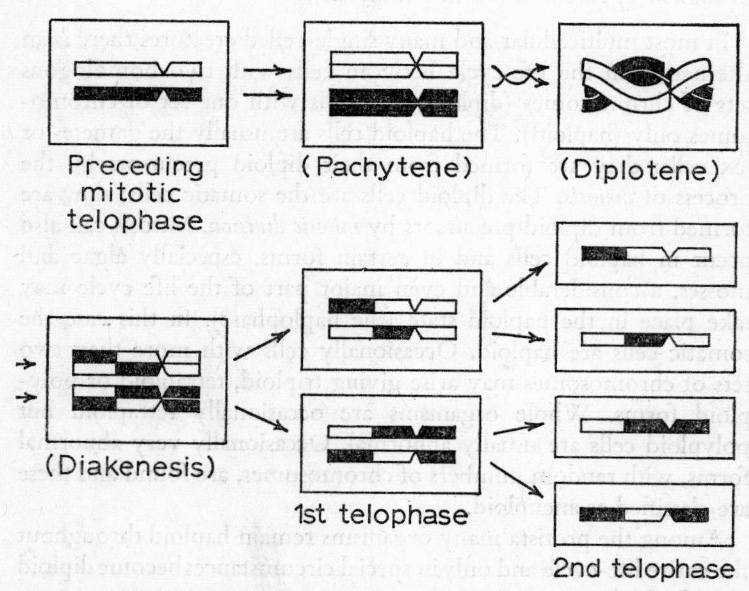

Figure 8.3. **The behaviour of a pair of chromosomes during meiosis.**

Mitosis and meiosis [233, 313, 314, 315, 452, 453]. Nearly all cells of higher organisms have a normal complement of two complete sets of chromosomes, that is, they are diploid. In the course of mitosis (see figure 8.2) each chromosome is duplicated and in this way two pairs of chromosomes are formed. When the cell divides one of each pair goes to each daughter cell which therefore ends up with exactly the same complement as its parent.

Meiosis (figure 8.3) is a special process which achieves a reduction of chromosomes from the diploid to the haploid number and is usually involved in the formation of sex cells. When two sex cells fuse to form a zygote the normal diploid complement is restored. Before entering meiosis each cell has two complete sets of chromosome pairs (as in ordinary somatic cells). The haploid chromosome number is attained as a result of two consecutive divisions, analogous in some respects to mitotic divisions except that they follow each other immediately and there is no doubling of the chromosomes between them.

Meiosis differs from mitosis in two other important respects. In the first place, during the first reduction division there regularly occurs an exchange of parts between the chromatids arising from two homologous chromosomes. This process of *crossing over* is of great importance since it results in each sex cell having a different combination of genes. The other major difference between mitosis and meiosis concerns the segregation of the chromatids. In preparation for mitosis each chromosome splits longitudinally into two chromatids which separate to form two daughter chromosomes, one of which goes to each daughter cell. In contrast, in the first meiotic division the chromatids remain attached at the centromere and when they separate at metaphase each pair of chromatids goes to the same daughter cell. In the second meiotic division the centromere splits (as in mitosis) and the two chromatids become independent chromosomes, one of each going to each daughter cell.

The end results of mitosis and meiosis thus differ in two major respects. In mitosis the number of chromosomes at the end of each cycle is exactly the same as at the beginning, whereas in meiosis the number of chromosomes in the daughter cells is only half that of the parents. Also the mechanism of mitosis tends to conserve the parent genetic complement in each daughter cell (although accidents occasionally result in differences), while in meiosis the mechanism specifically results in a redistribution of genetic material within each chromosome and each cell. The importance of this will become apparent later.

Little is known about the underlying mechanisms of cell division. The centrioles in animal cells seem to play an important part since they divide before all the other parts of the cell and then migrate to opposite poles with the initiation of division. From the centrioles radiates the spindle which probably takes the form of aligned, possibly contractile, protein molecules [10]. ATPase activity is associated with it. Each chromosome is attached to the spindle at the centromere and, on the addition of ATP to glycerol-fixed cells in metaphase, the chromosomes move apart. If the chromosomes are not properly attached, if the spindle is not properly formed or if the centrioles do not divide normally, then abnormalities of cell division follow. Although centrioles are essential for division in animal cells they are absent in plant cells.

Principles of genetic analysis

All genetic analyses depend on studying the orderly distribution of genetic factors during reproduction. From this the arrangement of the genes can be deduced.

The principle of segregation (Mendel's first law). Mendel found that pure strains of plants obtained by self-fertilization had certain stable characteristics (e.g. colour). When two plants with different characteristics were crossed, some of these features failed to appear in the progeny (F_1 generation). However, on recrossing F_1 individuals, the original characteristics always turned up again in an orderly fashion in the next (F_2) generation. The classical result obtained by Mendel is summarized in figure 8.4. These results can be explained by postulating two alternative genes for each characteristrc, one gene being dominant over the other when they both occur in the same cell. Two genes which behave in this manner occupy the same locus on corresponding (homologous) chromosomes (one from each of the two sets in a diploid organism), and are called alleles. The grouping of genes in the progeny of a cross (as in the above example) is called *segregation*.

If the segregation of two (or more) pairs of alleles is studied simultaneously they may behave according to Mendel's second

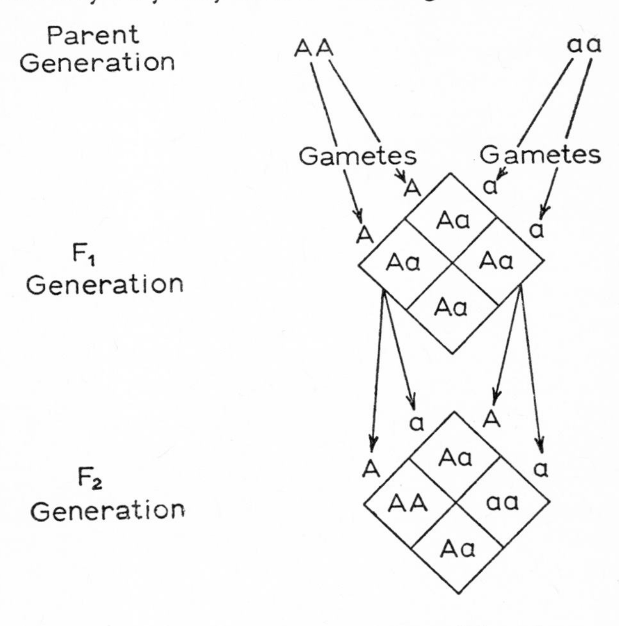

Figure **8.4. The principle of Mendel's first Law (the law of segregation).**

law, the law of independent assortment (figure 8.5). This law applies only if the two loci segregate independently and randomly, that is, if they occur on different chromosomes. Exceptions to this law were soon discovered and are of much greater importance than examples which conform since they provide evidence for *linkage* of genes. Genes are said to be linked when they segregate together with a much higher frequency than would be expected by chance. When genes occur on the same chromosome they are linked. A group of genes which segregates together is called a linkage group. All the genes in any chromosome form a linkage group.

Crossing-over and recombination. Linkage groups are not, however, immutable. For instance, chromosome breakage or fusion may alter

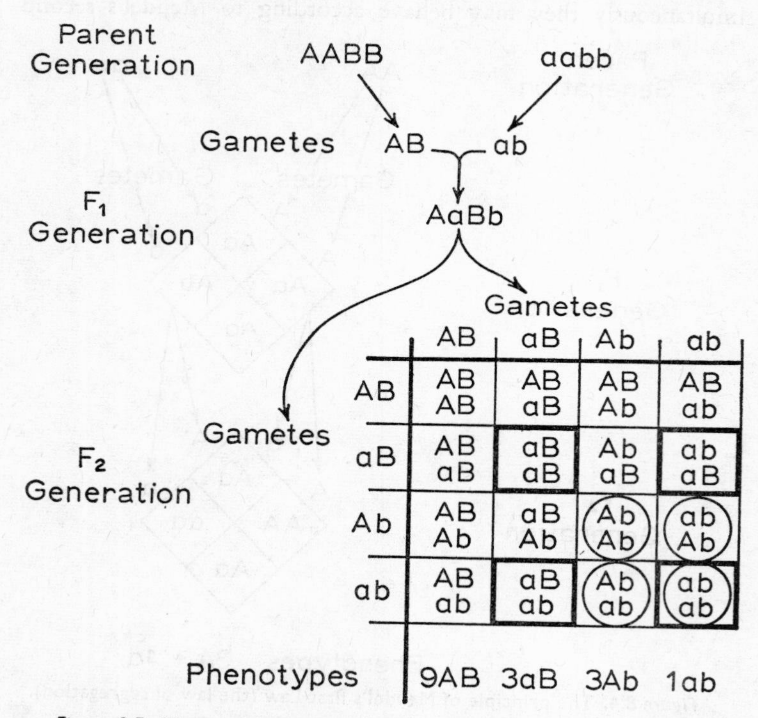

Figure 8.5. **The principle of Mendel's second Law (the law of independent assortment).**

the frequency with which groups of genes segregate together. The most important and commonest mechanism causing a rearrangement of genes in a linkage group is *crossing-over* (especially in meiosis). Crossing-over usually occurs between segments of homologous chromosomes. During the first meiotic division in diploid organisms the exchange takes place between adjacent chromatids (figure 8.3).

Crossing-over can also occur in mitosis in diploid organisms and during division in haploid organisms [389]; but, of course, since normally the daughter chromosomes are identical in haploid organisms, crossing-over is detectable only in special circumstances (to be described later).

The result of crossing-over is that genes in one linkage group are exchanged with genes in a homologous linkage group (figure 8.6). This reorganization of genetic material is called *recombination*. It is readily appreciated that two genes which are adjacent to each other

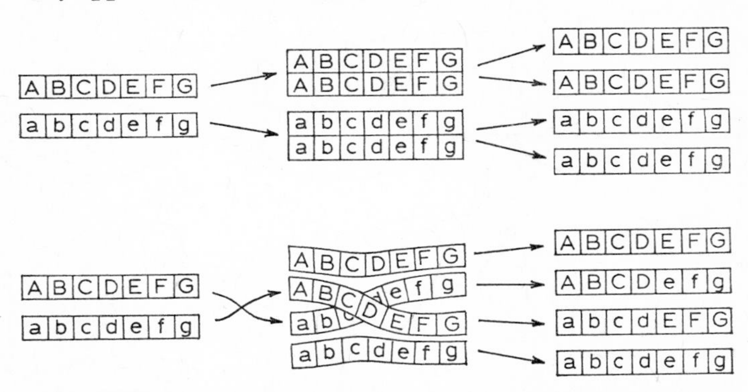

Figure 8.6. **The effect of crossing-over. In the upper figure two chromosomes, each with seven allelic genes, divide without crossing over, giving two pairs of daughter chromosomes, each identical with its parent. In the lower diagram crossing-over occurs giving rise to four different daughter chromosomes. These are recombinants.**

on a chromatid usually segregate together, whereas if they are widely separated there is a very good chance of their segregating separately. The frequency with which pairs of linked genes appear in new combinations is called the *recombination frequency* and it can be used as a direct measure of proximity of genes in a linkage group. By means of recombination analysis it is therefore possible to plot the relative positions of genes on a chromosome and hence to construct 'chromosome maps'.

Microbial genetics

Bacteria, being haploid, divide by simple fission and the progeny, barring accidents, have exactly the same genetic complement as

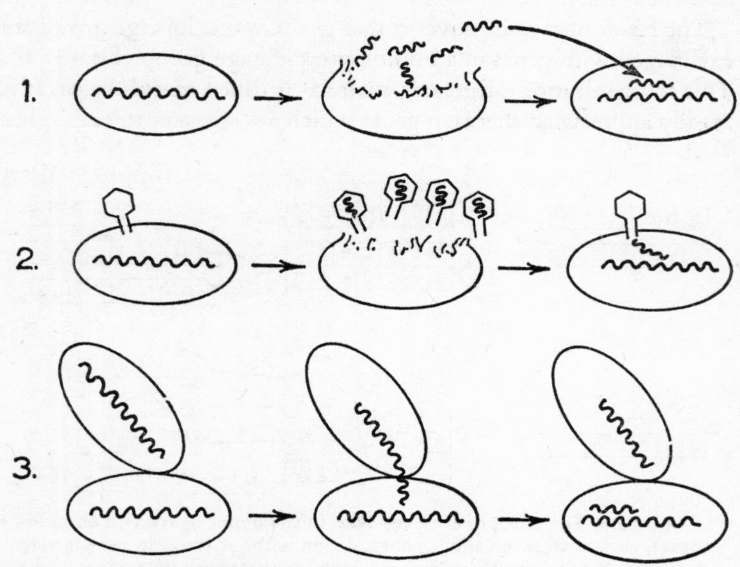

Figure 8.7. The processes which result in recombination in haploid micro-organisms. 1. Transformation. DNA from disrupted bacteria enters intact bacteria and is incorporated. 2. Transduction. When cells are killed by bacteriophage some bacterial DNA may be incorporated into bacteriophage DNA and carried with it to other bacteria. 3. Mating. This does not occur regularly in bacteria but only in the presence of a special gene.

their parents. There is normally, therefore, no possibility of recombination. There are, however, some special mechanisms whereby recombination may occur (figure 8.7). These are transformation, transduction and sexual (or parasexual) reproduction.

Transformation is a purely artificial process (figures 6.2 and 8.7) although something similar may rarely occur naturally. Genetic material (DNA) is extracted from one type of bacterium and used to treat other bacteria which may then be shown to have acquired some of the characteristics of the donor. Its main importance is in providing evidence that DNA is the genetic material.

Transduction is a similar phenomenon but in this case genetic material is conveyed from one kind of bacterial cell to another by means of a bacterial virus called a *bacteriophage* (or *phage*) [210, 529]. A typical bacteriophage (figure 8.8) is a complex structure, bearing a superficial resemblance to a hypodermic syringe. The 'head' portion contains DNA and the tail corresponds to a hypodermic needle, through which the DNA can be injected into a bacterium. The tail has an outer contractile sheath connected to fine fibrils which make the first contact with the coat of the bacterium and partly digest it by virtue of the enzyme, lysozyme, which is

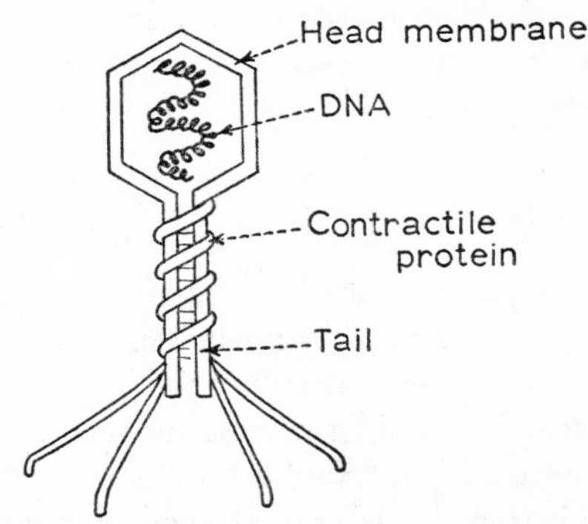

Figure 8.8. The structure of T4 bacteriophage.

associated with them. The outer sheath contracts, plunging the inner core into the bacterium and the DNA from the head is passed into the bacterial cytoplasm. The virus DNA then takes over the direction of the cell's metabolism from the host's DNA and diverts it to the synthesis of viral DNA and protein. When many new virus particles have been synthesized the infected bacterium ruptures, permitting them to escape and seek new victims. Sometimes some of the host DNA may be incorporated intact into some of the new virus and carried with it to a new host. If the virus enters but fails to kill the new host (that is, becomes latent) the host may then exhibit some of the characteristics of the old host.

Sexual (or parasexual) reproduction can also occur in bacteria and provides the main opportunity for genetic recombination. During this process bacteria may behave as females (recipients) or as males (donors). The donors behave in a manner similar to that described for bacteriophage, in that the male comes into contact with the female and then proceeds to inject its entire genome (complement of genetic material) into the female. The donor chromosome and the recipient chromosome come to lie together and crossing-over may take place between them. During this phase, until it divides, the organism is therefore briefly diploid. An interesting feature of this process is that the chromosome from the donor enters the recipient in an orderly manner and during the course of several minutes. It is therefore possible, by arresting the process before it is complete, to arrange that only part of the chromosome is transferred. By arresting entry of the donor chromosome at different times the order of genes on the chromosome can be determined.

Bacteriophage genetics

Bacteriophages (often shortened to phage) obey very simple genetic laws and the information they provide is of very great importance in connection with the nature of the genetic code [34, 35]. They must therefore be considered in some detail.

If a colony of bacteria is infected with phage the entire life-cycle is completed within minutes. If a pure inoculum of phage is used the progeny are all identical except for mutants, which arise randomly at the rate of about one mutation in each gene in every 100,000 to 1,000,000 phage. If a colony of bacteria is inoculated with two kinds of phage then a proportion of the bacteria will be infected with both simultaneously. If the different phage DNA strands are sufficiently homologous then crossing-over and recombination occur freely between them, within the bacterium, and many of the phage progeny are therefore *recombinants*.

Phage are commonly studied by inoculating a very small number of phage particles (say, a hundred) on to agar culture plates bearing confluent cultures of bacteria (sometimes called a 'lawn' of bacteria). Where each particle lands it attacks a bacterium which is killed and, on bursting open (lysis), infects neighbouring bacteria. In a matter of hours a clear area of dead bacteria, called a 'plaque', can be seen at each site of infection. Certain mutants can be readily recognized by the shape of the plaque they produce and also by the kinds of bacteria they attack. The susceptibility of certain bacteria also provides a method for selecting out mutants. For example, if a mutant, present as only one in a million phage, is capable of lysing a strain of bacteria which the other members of the population cannot attack, then it can easily be isolated by inoculating a few million phage on to a lawn of bacteria of this kind. The few plaques which occur will be derived entirely from the mutant.

Information of the utmost importance has been obtained by mapping mutational sites on the genome of one particular bacteriophage, the T4 phage. The DNA of this phage has about 200,000 base pairs, sufficient to carry information for 60,000 to 70,000 amino acids. Most investigations have been concentrated on one particular area of the viral genome – the so-called rII locus which forms only a few per cent of the whole. Phage with mutations at the rII locus can be readily recognized by two characteristic features. (i) On strain B of *E. coli* they produce characteristic large irregular plaques (r plaques). (ii) They fail to grow on strain K12 of *E. coli*, whereas the wild strain grows as well on it as on strain B.

Failure of rII mutants to grow on K12 is apparently due to a deficiency which can be compensated for if the wild strain is present simultaneously. If strain K12 is treated simultaneously with wild strain and an rII mutant, some bacteria will be doubly infected and these will produce both wild-type and rII mutant phage. A few recombinants between the two will also arise as a result of crossing-over and this forms the basis of recombination analysis of the rII locus of T4 phage.

One of the immediate results of applying fine genetic analysis to this locus was the definition of the *cistron* by means of the *cis-trans* or complementation test [34, 35]. The cistron is regarded as the functional genetic unit and is thought of as the stretch of DNA which codes for a single polypeptide chain. (There is a tendency to use the term interchangeably with 'gene' in current writing.) If a mutation occurs in a cistron, the corresponding polypeptide is non-functional. Consequently the organism does not make the protein

	Arrangement CISTRON 1 CISTRON 2	Result	Conclusion
Cis		Functional (One complete genome)	Mutations on different cistrons
Trans		Functional (One complete genome by complementation)	
Cis		Functional (One complete genome)	Mutations on same cistron
Trans		Non-functional (No complementation)	

Figure 8.9. The principle of the *cis-trans* complementarity test. As described in the text, this test depends on the fact that if a single bacterium is infected with two different mutant bacteriophage it will be lysed by them if the mutant loci are on different cistrons (genes) since complete recombinants can be formed by complementation. If mutant loci are on the same cistron, however, no complementation is possible since no recombinant can be formed in which a functional cistron is present.

parts of the phage. It therefore does not survive unless a functional allelic cistron is present simultaneously to code for the formation of the polypeptide. In phage this can happen if a bacterium is simultaneously infected with a complete phage as well as a defective mutant. This is the basis of the phenomenon described above, in which rII phage can grow in K12 bacteria if they are simultaneously infected with wild-type phage. It is also the basis of the *cis-trans* test. If K12 is infected, not with a mutant plus wild-type but with two mutants, then, if both mutations involve the same cistron, no complementarity is possible and the phage will not survive. On the other hand, if the mutations involve different cistrons both poly-

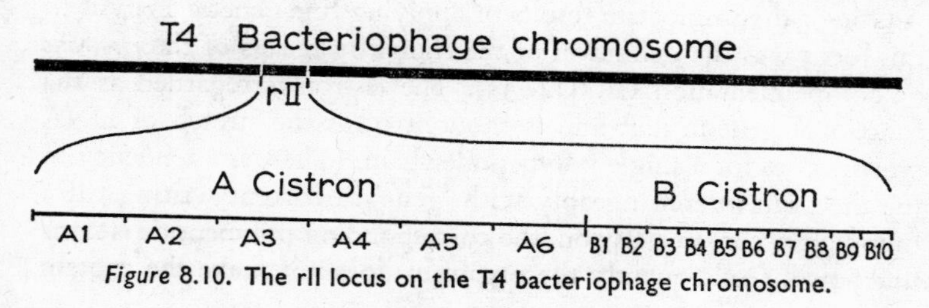

Figure 8.10. The rII locus on the T4 bacteriophage chromosome.

peptides will be formed; the two kinds of phage will therefore complement each other and both will survive.

The test is actually carried out as follows (figure 8.9). If the two mutants are classified as *X* and *Y* a recombinant is first made (*XY*). Four different forms of the phage are now available, namely, *X*, *Y*, *XY* and *O* (wild-type). The *cis* test (control) is made by inoculating K12 with *XY* and *O* simultaneously. Because *O* provides a complete set of cistrons plaques should form. The *trans* test is made by inoculating K12 with *X* and *Y* simultaneously. If the two mutations are on the same cistron no plaques will be formed, since there is no opportunity for complementation. On the other hand, if they are on different cistrons complementation will permit phage growth and plaque formation. By this test it has been found that the rII locus of T4 phage consists of two cistrons, called the *A* and *B* cistrons respectively. These have been further divided into segments of which there are 59 in the *A* cistron and 21 in *B*, numbered *B*1 to *B*21 (figure 8.10).

Mapping phage. In mapping phage the two techniques used are (i) the *cis-trans* test of complementarity and (ii) recombination between mutants and *deletions*. Deletions are variants in which a large part of the genome is completely deleted or inactivated. They never spontaneously revert to the wild type whereas point mutations frequently do. A large series of deletions has been collected and characterized, varying from complete deletion of both *A* and *B* cistrons to deletions of very small parts of either. In locating a mutation a series of recombinants is first made between the mutant and a series of deletions. If the mutation is in an area which is not affected by the deletion then functional recombinants can be obtained. However, if the mutation is in the area affected by the deletion then no complementation is possible. In this way it is possible to assign mutations to limited areas of the genome and the final details of mapping can be carried out by the *cis-trans* test, which will ultimately distinguish between nucleotide pairs. (The smallest unit of recombination has been called the *recon*.)

Mutagens [152]. Many mutants arise spontaneously, but the spontaneous mutation rate is very low. It can be increased greatly by many treatments, for example irradiation. Substances which induce mutation are called mutagens; most are known to act on DNA. For instance nitrous acid deaminates cytosine to uracil in DNA. Also, a chemical analogue, such as 5-bromo-uracil, can replace a normal base (in this case, thymine), and give rise to errors in subsequent replication of the DNA molecule. Mutants caused by substitution of one base by another do not, as a rule, completely lose the affected function, but they show an alteration or a great diminution of it. (They are referred to as 'leaky'.) Certain other mutagens give rise to non-leaky mutants, that is, mutants in which the affected function is completely lost. Principal among these are the acridine dyes (such as proflavine), which produce mutations by causing an extra base to be inserted into the DNA chain [247]. Mutants of this kind were used in providing evidence for the nature of the genetic code.

The genetic code

The principle of colinearity between nucleic acids and proteins was proposed some time ago [96]. It is implicit in this idea that the structure of DNA somehow determines the sequence of amino acids in proteins and polypeptides. Some of the strongest pieces of evidence in support of it have come from studies of amino acid sequences in proteins which differ from each other as a result of mutations. For instance, the difference between normal haemoglobin and the haemoglobin of patients with sickle-cell anaemia is simply that one glutamic acid residue in normal haemoglobin is replaced by valine in sickle-cell haemoglobin [242, 243, 244]. In another abnormal haemoglobin the same amino acid is replaced by lysine. It was therefore postulated that the sequence of nucleotides in DNA might determine the order of amino acids in proteins. Since there are only four bases but about twenty amino acids numerous codes were proposed, many postulating units comprising triplets of three nucleotides. There are 64 possible combinations of three items out of four and this is an adequate number. This idea has now proved to be correct although in its original conception it was no more than a speculation.

Two general types of linear code are possible. In overlapping triplet codes, for instance, the first, second and third symbols code for one message, the second, third and fourth for a second and so on. In a non-overlapping code the first, second and third symbols again code for the first message but the second message would be coded by the fourth, fifth and sixth symbols. Had the genetic code been of the overlapping type it would have been predicted that changes in single DNA bases would result in changes in more than one amino acid. However, by producing mutations in tobacco mosaic virus by nitrous acid it was shown that this was not the case. Hence a non-overlapping code seemed more likely.

Sequence	
CAT | CAT | CAT | CAT | CAT	Normal sequence
CAT | **G**CA | TCA | TCA | TCA | T	+ mutation (Extra base)
CAT | CAT | C:TC | ATC | AT	– mutation (Base missing)
CAT | **G**CA | TC:T | CAT | CAT	± recombinant

Figure 8.11. The basis of the experiment by Crick *et al.* which provides evidence for the general nature of the genetic code. Insertion of an extra base moves the message out of register so that it no longer makes sense. Deletion of a base does the same. If a base is added and another is deleted nearby most of the message makes sense again.

Genetic evidence for the general nature of the genetic code was obtained by Crick and his colleagues [97, 98], employing acridine-induced mutations of the *B*1 segment of the *B* cistron of the rII locus of T4 phage. One mutant, called FCO, was used as the basis of the experiment. Suppressors of this mutant were also isolated (that is, mutants which, when recombined with FCO, gave a wild type – or more properly a pseudo wild-type – phage). Since acridine mutants have an additional base [289] (hereafter designated as +) it was suggested that the suppressors were mutants which had lost a base (designated as —). In agreement with this they were found to be non-leaky. The explanation of the effect of the suppressor then was as follows. It was assumed that in the wild type the code was read in triplets. When an extra base was inserted the triplets no longer made sense (figure 8.11). Removal of a base a little further along the chain as a result of recombination with a suppressor led to the establishment of the correct sequences once more. Consequently if the mutations were situated reasonably near each other a functional polypeptide might be formed. It would be surprising, on this theory, if sense could be restored when the + and − mutant loci were some distance from each other on the cistron, and this is what was found.

This system now provided an opportunity to test for the general nature of the code, assuming it to be of the nature of a regular non-overlapping type. Double and triple mutants were prepared by recombination and it was found that several of the triple mutants (+ with + with + or − with − with −) were fully functional. Consequently, the code seems to be of the non-overlapping triplet type read in a regular manner from one end to the other, since it requires three single moves of the register to restore sense in the transcription.

Independent evidence for the triplet nature of the code has come from biochemical studies which have also led to elucidation of its details. Nirenberg and Matthei [312, 345] found that polyuridylic acid added to a system of ribosomes, ATP and enzymes promoted the incorporation of ^{14}C-phenylalanine into polypeptide material, which turned out to be polyphenylalanine. This experiment suggested that a series of uridylic acid residues in RNA coded for phenylalanine and that, on the basis of a triplet code, the triplet for phenylalanine would be UUU. Nirenberg's [253, 310, 311] and Ochoa's [23, 157, 287, 288, 445, 446, 487, 488] groups exploited this tool by preparing many kinds of polynucleotides of known composition and adding them to a similar system. From the products formed it proved possible to deduce the bases, forming the triplets which code for each amino acid.

The groups of workers employing genetic and biochemical methods both predicted at an early stage that the code would prove to be degenerate, that is, that individual amino acids would be coded for by more than one triplet. This has turned out to be so. Two or three triplets for each amino acid have now been found.

The first experiments could tell only which three nucleotides coded for each amino acid. Nothing was known about the order in which they occurred. For example, it was known that 2U, 1A coded for tyrosine but the correct sequence could have been AUU, UAU or UUA. The problem has been solved by Nirenberg's group who found that trinucleotides of known structure (for instance, pUpUpG) would cause specific sRNA molecules, with their attached amino acids, to be bound to ribosomes, even in the absence of protein synthesis. By using this as a tool the sequences of a large number of triplets have been determined [36, 280, 281, 344].

Further conclusive evidence for the composition and polarity of certain triplets has come from the laboratory of Khorana who prepared polynucleotides with regularly repeating sequences of the same two or three bases and used these as templates for polypeptide synthesis [252, 346, 347].

The sequences for the other code triplets have been deduced by comparing the amino acid changes caused by simple mutations, which were assumed to result from a single base-change. For instance, it was known that tyrosine (UAU) was sometimes replaced by cysteine which was known to be coded for by 2Us and 1G. It was therefore deduced that the most likely event was a change from A to G; hence the sequence of triplets in cysteine is likely to be UGU. By comparing a very large number of mutations of this kind the sequences for all known triplets have been deduced.

The most elegant example of this kind of study comes from Yanofsky's work [188, 522]. He and his colleagues have isolated many mutants affecting the same amino acid residue of the enzyme tryptophan synthetase. By analysis of the enzyme protein the substituent amino acids have all been identified and their relationships with each other compared with the triplet assignments as shown in figure 8.12.

The code as at present envisaged is shown in table 2 [254, 255, 344]. It is noteworthy that some general rules seem to emerge. One is that the possible triplets for each amino acid always have the first two bases in common. The third base can vary. In some cases it can be U, C, A or G but in most cases it can be only one of the pair U and C or the pair A and G. These relationships probably have to

***Table 2.* The Genetic Code (1965)**

Amino acid	Code triplets (codons)				
	Common bases	Complete triplets			
Phenylalanine	UU–	UU	UUC		
Leucine	UU–			UUA	UUG
	CU–	CUU	CUC	CUA	CUG
Isoleucine	AU–	AUU	AUC	AUA	
Methionine	AU–				AUG
Valine	GU–	GUU	GUC	GUA	GUG
Serine (1)	UC–	UCU	UCC	UCA	UCG
Proline	CC–	CCU	CCC	CCA	CCG
Threonine	AC–	ACU	ACC	ACA	ACG
Alanine	GC–	GCU	GCC	GCA	GCG
Tyrosine	UA–	UAU	UAC		
End of message	UA–			UAA	UAG
Histidine	CA–	CAU	CAC		
Glutamine	CA–			CAA	CAG
Asparagine	AA–	AAU	AAC		
Lysine	AA–			AAA	AAG
Aspartic acid	GA–	GAU	GAC		
Glutamic acid	GA–			GAA	GAG
Cysteine	UG–	UGU	UGC		
Tryptophan	UG–				UGG
Arginine	CG–	CGU	CGC	CGA	CGG
	AG–			AGA	AGG
Serine (2)	AG–	AGU	AGC		
Glycine	GG–	GGU	GGC	GGA	GGG

do with the way the 'anticodon' triplet in sRNA recognizes the 'codon' triplet in mRNA.

The 'end of message' triplets, which signify the termination of a peptide sequence, have been deduced from genetic studies using

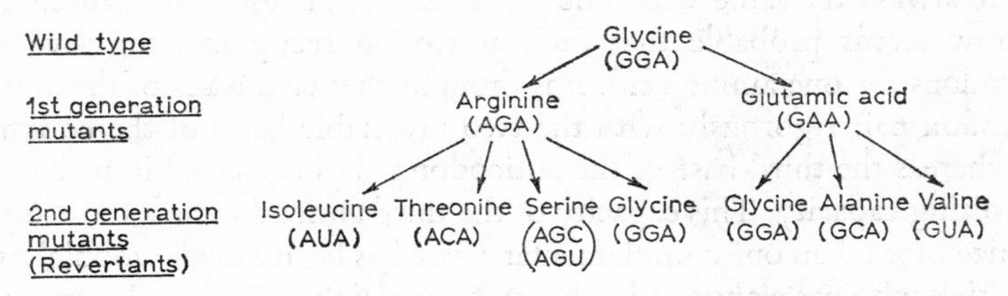

Figure **8.12. Mutations found by Yanofsky in tryptophan synthetase. The first-generation mutants could not synthesize tryptophan; the amino acid changes shown were found at one residue only in the entire molecule. The second-generation mutants reverted, i.e. recovered the ability to synthesize tryptophan completely or at a reduced rate. New substitutions at the same amino acid residue were found. Each single-step mutation was compatible with a single nucleotide change as predicted by the genetic code.**

special mutants of bacteriophage (called 'amber' and 'ochre' mutants) [60] in which premature termination of the synthesis of some proteins occurs. This is due to substitution of a base in a triplet specifying an amino acid, which becomes converted to a triplet specifying 'end of message' as a result. By a study of prematurely terminated polypeptides from mutants of this kind, and comparison with the whole polypeptides from wild-type organisms, it has proved possible to identify the triplets which have been affected by mutation. By this means and also by some rather complicated genetic studies the end of message triplets have been deduced [59, 60, 63, 498].

Anticodons and wobble

The genetic code has been described as it occurs in messenger RNA. In the translation of this message into a polypeptide sequence each nucleotide triplet (or codon) becomes associated with a tRNA molecule, as described in chapter 6. The site on the tRNA molecule which recognizes the codon is thought to be a triplet of bases (referred to as the anticodon) complementary to the three bases of the codon.

It has become clear that for nearly every amino acid there are several codons, generally related by the fact that the first two bases are always the same while the third may be one of two or three. It now seems probable that each anticodon recognizes the several codons for one amino acid. It is thought that two bases of the anticodon pair rigorously with the two invariable bases of the codon, whereas the third base of the anticodon is less restricted in its base-pairing capacity. This capacity of the third anticodon base to recognize more than one complementary base has been called 'wobble' by Crick who has elaborated a theory to explain it. The consequence is that each tRNA recognizes several codons for each amino acid. Hence, the number of tRNA molecules necessary for the translation of the genetic code is considerably less than the number of codons. Now that the nucleotide sequences of several tRNA molecules are known some anticodons have been provisionally identified; the third base in some cases has proved to be, not one of the common four bases, but one of the unusual bases (table 3).

Table 3. Probable codons and anticodons for four amino acids

Amino acid	*Alanine*	*Serine*	*Tyrosine*	*Valine*
Codon (5'→3')	U GC C A	U UC C A	U UAC	U GUC A
Anticodon (3'←5')	CGI	AGI	AψG	CAI

(I – inosine; ψ – pseudouridine)

9: Interrelationships among Intracellular Structures

In the structurally simplest cells, such as bacteria, there is only one compartment, enclosed by the cell membrane. In this compartment chromosomes, ribosomes and unbound enzymes probably float freely. Such proteins as are bound are attached to the cell membrane or to other insoluble structures such as chromosomes or ribosomes. There are no structures like mitochondria: the enzymes of respiratory metabolism, if present, are bound to the cell membrane (where they may be aligned as they are in the mitochondria of more complex cells). In simple cells of this type, in which the greatest dimension may be of the order of 1 μ, small molecules probably pass from one site to another by simple diffusion; no special mechanisms are necessary to connect different structures other than the enzymic control systems described in chapter 7.

In some of the more complex Protista, especially the Protozoa, and in the cells of Metazoa and Metaphyta, the situation is very different. They are generally larger, some being several centimetres long. The cytoplasm is divided into compartments by the membranes of the endoplasmic reticulum and Golgi apparatus, and special functions are localized in membrane-enclosed organelles, such as mitochondria, plastids and the nucleus. Some of these structures have a high degree of autonomy, but clearly, for the efficient working of the whole cell, their functions must be rigorously co-ordinated.

Topological relationships

Membranes are by no means static structures. During mitosis the nuclear membrane completely disintegrates in a few minutes; later it is reconstituted equally quickly. A mitochondrion is a highly flexible structure and may be seen, in time-lapse films, to change its shape and size continuously; hence the membranes which form its skeleton must also be in a continuously changing state. Cell membranes and the membranes of the Golgi apparatus and endoplasmic reticulum can also fragment and reform. This provides a means of communication between different compartments. The key to this arrangement is that the extracellular environment, the contents of the Golgi cisternae, the cisternae of the endoplasmic recticulum and of lysosomes are topologically equivalent, while the cell sap, in which the nuclear contents, mitochondria and plastids float, is always separated from them (figure 9.2). Vesicles arising from the Golgi apparatus, the endoplasmic reticulum, lysosomes or the cell surface can apparently fuse again with any other of these structures. Hence, the cisternae of the Golgi apparatus, the endoplasmic reticulum and lysosomes can probably communicate with each other and the extracellular environment by this process. Sometimes, direct connections, e.g. of the cell surface with the endoplasmic recticulum, can be directly observed.

These related structures provide mechanisms for excretion, digestion and transport in cells. In pancreatic acinar cells, protein is first secreted directly into the cisternae of the endoplasmic reticulum (figure 9.1 (1)); this communicates with the Golgi apparatus, either directly or by vesicle formation and fusion. Thus zymogen accumulates in the Golgi apparatus, where it is concentrated and forms granules. Vesicles containing zymogen granules then bud off and migrate to the cell surface, where they fuse and extrude their granules [78].

Material is ingested by a reversal of this process (figure 9.1 (2)). Invaginations of the cell membrane give rise to vesicles containing extracellular material. These pinocytic or phagocytic vacuoles move into the interior of the cell. (Sometimes, as in *Paramecium*, they follow a well-defined track.) Water is absorbed from them and they

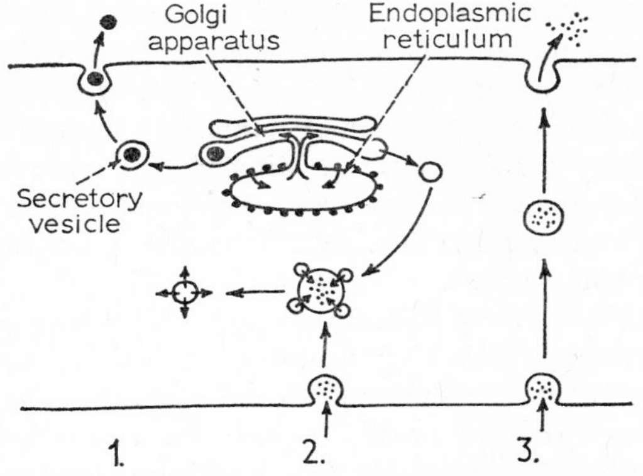

Figure **9.1. The relationships between secretion (1), pinocytosis and phagocytosis (2) and transcellular transport (3). For description see text.**

become smaller. Simultaneously, small bodies, which have been called protolysosomes and are thought to contain digestive enzymes, fuse with them and secrete their contents into them, so that any digestible matter is reduced to small components which can diffuse out into the cell sap [57, 173, 248]. The resulting small vacuoles are

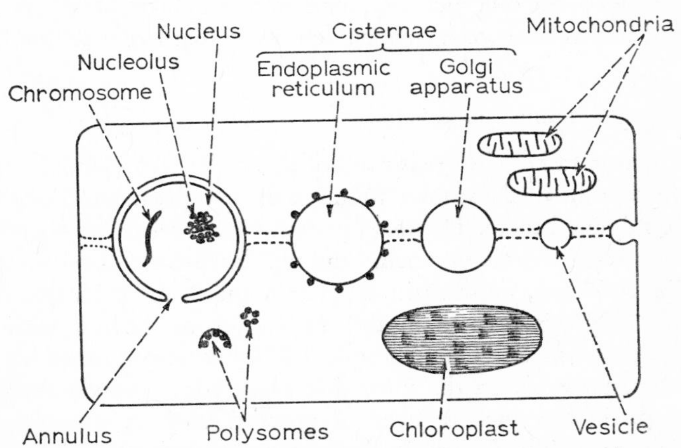

Figure **9.2. Diagrammatic representations of a cell, showing the topological relationships among subcellular structures.**

called lysosomes and may correspond to digestive vacuoles in Protozoa. The protolysosomes may originate from the Golgi apparatus by the budding off of vesicles in the same way as excretory vesicles containing zymogen. Digestion of cytoplasmic organelles probably takes place by a similar process. Mitochondria may, for instance, become incorporated in digestive vacuoles and so give rise to the structures which have been called cytolysosomes, autolysosomes or autophagosomes [57].

Invagination of the cell membrane to form vacuoles is, in certain cells, directly linked with a process of refusion of the vacuoles with the cell membrane elsewhere and discharge of their contents (figure 9.1 (3)). Such cells are usually orientated within a cell layer; invagination occurs at one end of the cell while extrusion occurs at the opposite end, so that extracellular material is moved from one end to the other. In endothelial cells of blood vessels and the kidney and the epithelial cells of the gut this mechanism provides for the active transport of materials across a cell layer.

The cell, then, is divided into two major compartments. One, which we may call the cisternal complex, is functionally continuous with the environment. The other is the cell sap: it comprises the nuclear sap and the cytoplasmic sap which may be continuous. The organelles of the cell, such as mitochondria and protoplasts, are contained in this latter compartment and may float freely within it.

Nucleocytoplasmic connections

It is now rather clear that most of the nucleic acids of the cell are formed in the nucleus; those found in the cytoplasm are thought to have migrated there. The main exceptions are related to the DNA of some cytoplasmic organelles and will be discussed later.

The evidence, outlined in chapters 6 and 8, indicates that messenger RNA is formed in the nucleus, in association with the chromosomes, and that ribosomal RNA is also formed in the nucleus, probably in the nucleolus. Since some protein synthesis goes on in the nucleus (even of proteins, such as hæmoglobin, which are mainly formed in the cytoplasm), messenger RNA and ribosomes perhaps first become associated within the nucleus, to form functioning polysomes. Most polysomes are, however, found in the cytoplasm, and it is assumed that they migrate there through the pores in the nuclear membrane.

Recently, evidence has been obtained for the existence of a particle (with a sedimentation value of 45 S) which first appears in the nucleus but rapidly becomes associated with the polysomes in the cytoplasm [206, 249, 250, 279, 298]. This particle would seem to contain a precursor of the 40 S ribosomal subparticle and possibly mRNA. By association with a 60 S subparticle, either in the nucleus or the cytoplasm a nascent polysome, comprising a molecule of mRNA with one ribosome, may be formed. Other ribosomes may then become attached; it is quite likely that they are, in fact, extensively re-used in the cytoplasm.

Some polysomes remain free in the cell sap but become bound to the membranes of the endoplasmic reticulum. It appears to be the 60 S components of the ribosomes which associate with the membrane of the reticulum in such a way that the 40 S particles protrude into the cell sap. The 40 S particles which are the ones associated with messenger RNA are, therefore, directly accessible to new messenger RNA coming from the nucleus, whereas the new peptides formed are automatically excreted into the cisternae of the reticulum. In excretory cells the proteins in the cisternae may then be passed to the Golgi body and ultimately excreted in the manner already described (figure 9.1).

Independence and interdependence of intracellular organelles

According to the scheme outlined above cytoplasmic nucleic acids are synthesized in the nucleus. There are two other possible ways in which they might originate. For one thing, RNA might act as a template for synthesis of more RNA in the cytoplasm; and, for another, the cytoplasm might have DNA-containing structures. In infections with RNA-containing viruses, RNA-primed RNA synthesis does occur in cells; but in the normal cell it almost certainly does not because the necessary enzyme has not been found. We shall therefore not consider further the first possibility. On the other hand, in recent years much evidence has accumulated concerning cytoplasmic DNA.

Cytoplasmic DNA falls into three main categories. In some egg cells, for example, those of amphibia, there is a very large amount of DNA in the cytoplasm, sometimes of the order of a thousand-fold excess over the DNA of the pronucleus. The function of this DNA is completely unknown. It may merely provide a pool of material which can be degraded to give molecules from which new DNA can be formed. Or it may have a more interesting function in regulating the early development of the embryo. This problem has not been resolved.

Secondly, many cells, especially of microorganisms and Protozoa [202, 414], have self-replicating DNA-containing particles, almost indistinguishable from vegetative virus particles, called episomes. Episomes are self-replicating particles which may exist free in the cytoplasm of the cell or, alternatively, in intimate association with the chromosome. The classical example of an episome is the mating factor (F) in bacteria. This endows bacteria with the property of conjugating with other bacteria. Bacteria containing the mating factor may conjugate with other bacteria which lack it, and infect the F^- cells. When the mating factor is integrated into the bacterial chromosome it can still promote mating with F^- bacteria, but the bacterial chromosome is then transferred to the recipient along with the mating factor. Episomes not only provide the information needed for their own replication but they also give rise to products which influence the behaviour of the cells containing them. They may repress transcription from parts of the genome of the host cell or they may code for enzymes and other proteins which participate in the metabolism of the cell.

Thirdly, many intracellular organelles contain DNA and there is evidence that this is functional. For example, DNA occurs widely, and probably universally, in mitochondria of cells of higher organisms [338, 339, 422]; it also occurs in plastids [406, 407, 415] and the basal bodies of cilia. This DNA is demonstrably different from nuclear DNA. Moreover, inheritance of certain characteristics of mitochondria and plastids may be independent of chromosomal inheritance in the cell. Besides chloroplasts, some strains of yeast have leucoplasts – plastids which do not contain chlorophyll. In these the inheritance of leucoplasts and chloroplasts does not follow the usual roles of segregation; the mutant 'gene' determining that leucoplasts will not synthesize chlorophyll may, therefore, reside in the plastid itself. Certain mutants involving mitochondrial function ('petite' and 'poky' mutants of yeast and Neurospora) behave similarly. Hence, some intracellular organelles are in part genetically independent of their host. However, certain functions of these organelles are certainly controlled by the nucleus: for example, the structural gene for cytochrome c behaves as a normal nuclear gene. The genetic control of mitochondrial and plastid functions may, therefore, be shared between the RNA of the whole cell and the DNA of the organelle.

Compartmentation [65]

Evidently, structures like mitochondria and chloroplasts have quite a high degree of functional autonomy. As has already been discussed, in cells which have chloroplasts, photosynthesis is confined to them; similarly, where there are mitochondria, oxidative phosphorylation occurs only in them. Moreover, many of the small molecules that play a part in the metabolic processes in these organelles, such as ATP, NAD and NADP, are sequestered within them and do not exchange freely with other parts of the cell. Since one of the main functions of the mitochondria is to oxidize the end-products of glycolysis (which are formed in the cell sap), and to make ATP and reduced NAD for other cellular functions, many of which occur outside mitochondria, special chemical connections are necessary between the mitochondria and the cell sap. Some of these have been identified; they form so-called 'shuttles'. The best known is the 'glycerophosphate shuttle' [285] which serves to illustrate the general principle (figure 9.3).

Although many substances, including NAD and NADH, cannot pass freely from mitochondria to cell sap, certain others, including glycerophosphate, can. In the cell sap the shuttle operates by the formation of glycerophosphate from dihydroxyacetone phosphate by reduction with NADH (figure 9.3). This involves an input of energy into glycerophosphate which can pass freely into the mitochondria and thus acts as an energy carrier. Within the mitochondria another enzyme removes the hydrogen from the glycerophosphate, and in so doing converts an intramitochondrial flavoprotein to reduced flavoprotein. The flavoprotein is reoxidized through the electron transport system yielding two molecules of ATP for each

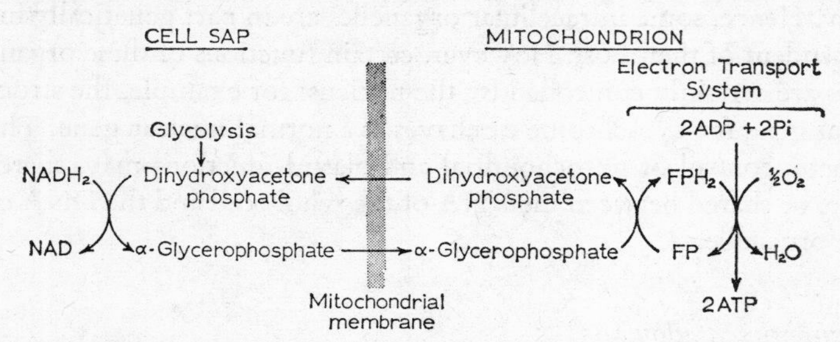

Figure **9.3. The glycerophosphate shuttle, showing how glycerophosphate acts as a hydrogen carrier from extramitochondrial NADH; to intramitochondrial flavoprotein (FP).**

electron pair; the dihydroxyacetone phosphate formed by reoxidation of the glycerophosphate is able to diffuse out through the mitochondrial membrane to complete the cycle. The final effect of this process is a reduction of intramitochondrial flavoprotein at the expense of $NADH_2$ from the cell sap, although the molecule of NAD never enters the mitochondrion. This amounts to a mechanism for transport of electrons which can be equated to an energy transport system.

The way in which high energy compounds formed in the mitochondria provide energy for reactions in the cell sap is less clear; but an enzyme, located in the mitochondrial wall, possibly has the function of transferring the terminal high energy phosphate group from intramitochondrial ATP to extramitochondrial ADP.

Similar mechanisms are almost certainly associated with the interchange between the chloroplast and the cell sap, but they have not been so intensively studied.

Intracellular movement

The cytoplasm of most cells is far from static and very active movements of intracellular structures can be observed. These are often rapid enough to be seen with the eye but, even where living cells seem relatively inert on ordinary examination, they can be shown by time-lapse cinemicrography to be the site of much activity. The principal movements are of chromosomes in mitosis and meiosis, changes in shape, size and position of mitochondria, streaming of the cytoplasm, rotation of the nucleus, expansion and shrinking of cytoplasmic vacuoles and the rapid motion of cytoplasmic granules.

Many of these movements have been thought of as passive but, with the exception of Brownian movement seen at high magnifications and the movement of granules in cytoplasmic currents, most of the others are almost certainly energy-requiring and often components of co-ordinated cell functions. The reason for saying this is that, with the exception of Brownian movement, the rates of all these other processes are highly temperature-dependent. Furthermore, in some glycerol-extracted cells, the addition of ATP re-initiates movements [11, 204, 218, 427].

Intracellular movements may be divided into those involving contraction and those involving flow.

Intracellular contractile movements are best exemplified by the behaviour of the spindle during mitosis. The spindle is made up of large numbers of very long, orientated fibrils with the appearance of microtubules about 200 Å in diameter. It is not certain whether these are themselves contractile or whether contractile protein is associated with them; currently there is prejudice in favour of regarding them as intrinsically contractile.

Microtubules of this kind are found associated with other cellular structures. They have no very regular distribution in animal cells, but in plant cells and some Protozoa they are packed under the cell membrane and parallel to it, to form a lining called the ectoplasm or cortical gel [282, 464]. When plant cells are disrupted, very small tubular structures are often released; these are highly mobile and can be seen to spin freely under the microscope. They are probably bundles of 200 Å microtubules, which therefore seem to possess some means of propelling fluid past them. This observation is of considerable significance in relation to cytoplasmic flow in plant cells since, as mentioned above, the microtubules are found just under the cell membrane and this is where cytoplasmic streaming is fastest. (It can be of the order of 60 μ per second.)

Another mechanism for cytoplasmic streaming in plants has been proposed by Ambrose and Goldacre [12]; they attribute it to electro-osmosis. This is the name given to the flow of water molecules which takes place over a charged surface when an electrical potential is established along it. Since the cell membrane carries static charges, all that is necessary to invoke electro-osmosis is an electrical potential along the membrane. These workers suggest that the sodium pump might provide areas of altered electrical potential on the cell surface such as would cause electro-osmosis. They have shown that strong electrical and magnetic fields affect cells in a manner which is predicted by this theory.

10: Cytodifferentiation

Morphogenesis implies the formation of the recognizably different parts of an organism. It includes two complementary but distinct phenomena, cytodifferentiation and tissue organization. These terms are usually applied to multicellular organisms and in this context cytodifferentiation implies the development from a single precursor of different kinds of cells, without any alteration of the genetic constitution. Tissue organization is the assembly of different cell types to form the tissues and organs of the body; it will be discussed in the next chapter. The definition of cytodifferentiation given above is in some ways restricted but it has the advantage that it pins us down to the consideration of a specific mechanism – the way in which genetic information is modified in the formation of a specialized cell.

Morphogenesis in protista

Much of the interest of cellular differentiation is in relation to the development of multicellular organisms but certain phenomena in unicellular organisms are very similar and have to be considered in the same context. Many of the Protista, possibly even some Bacteria [456], have highly specialized structures. Some of these organisms are of particular interest in providing information about the parts played by cellular components in determining cellular structures.

Acetabularia is a unicellular marine plant which grows to a length of several centimetres. The nucleus is situated at the base while at the other, free, end of the cell is a cap. The shape of the cap is characteristic of the species. It has been shown that the morphology of the cap is determined by the nucleus, which is easily removed by

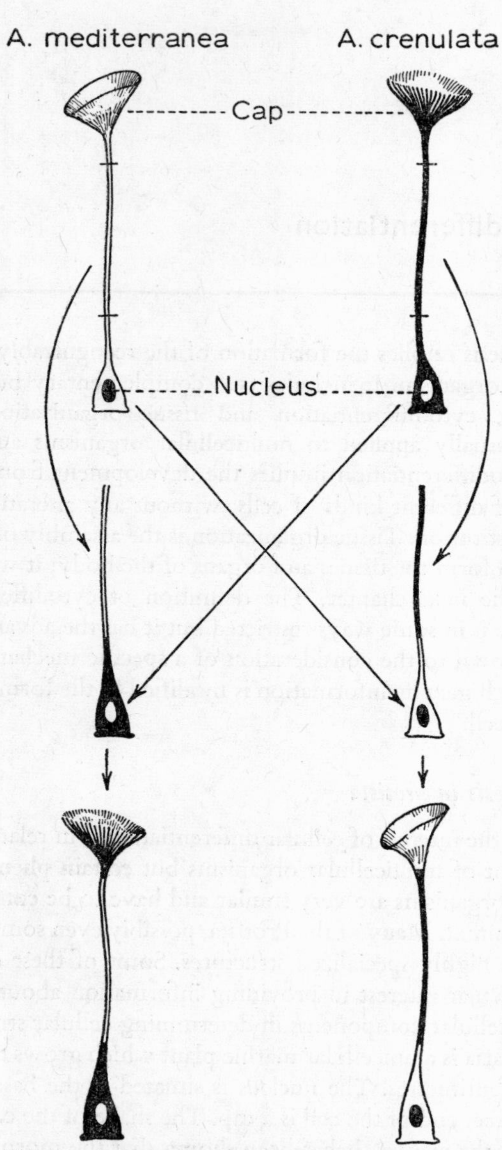

Figure **10.1. Transplantation experiments showing the influence of the nucleus on the regeneration of the cap in Acetabularia (after Hämmerling).**

amputating the base [193, 194]. It is possible to transplant nucleus and cytoplasm by grafting. When the nuclear end of *A. mediterranea* is attached to the decapitated cytoplasm of *A. crenulata* (figure 10.1) then the regenerated *Acetabularia* bears a *mediterranea* type cap. If the experiment is done in reverse, the regenerated organism again has the cap corresponding to the nucleus, in this case from *crenulata*. Evidently, in this organism, the nucleus is the main governing factor in determining the ultimate morphology of the cell. (The conclusion must be accepted with a little reservation since a considerable amount of cytoplasm is inevitably transplanted with the nucleus.)

In other protista, however, there is good evidence that the cytoplasm also plays an important independent part in the development of certain characteristics. This is sometimes due to the behaviour of certain cytoplasmic inclusions, such as the kappa particles in *Paramecium* [442] and the plastids in *Algae*, which have some DNA of their own and can behave in a genetically independent fashion. When these cases are eliminated, however, there are still a few examples where the evidence very strongly suggests that cytoplasmic factors by themselves may make an important contribution to the ultimate form of the cell. For instance, transplantation of a piece of cell wall, including the kinetie, in *Stentor*, may determine the morphology of the recipient area [462]. Also, streptomycin resistance in *Amoeba* and *Chlamydomonas* may depend on cytoplasmic rather than nuclear factors [87, 414]. These features of Protista have to be emphasized before considering multicellular animals, since they may have an important bearing on early embryonic development.

Some protozoa exhibit another aspect of behaviour of importance in relation to cyto-differentiation in higher animals, in that they can change their morphology profoundly during different stages of their life cycle. There seems little reason to doubt that throughout the entire cycle they possess exactly the same genetic material and yet they may vary from amoeboid to flagellate forms and may even form gametes. The *Plasmodiidae* (the organisms of malaria) are well-known examples of this behaviour. Clearly even at this level, very profound modifications of gene expression can occur.

The behaviour of certain organisms like the sponges (*Porifera*) and slime moulds (*Mycetozoa*) is a little more akin to that of true multicellular organisms. All the cells in a sponge are derived from the same original cell and they live more or less independent lives. A sponge is, therefore, a cell colony. On the other hand, it includes several recognizably different kinds of cells which behave in a specialized manner. The slime moulds are particularly interesting, since during part of their life cycle they exist as free unicellular organisms while, during the other part they form either an organized syncitium or a multicellular organism [47, 48, 271, 425].

Differentiation in multicellular organisms

Cytodifferentiation [146, 291]. In multicellular plants and animals, genetic continuity from one generation to another is maintained

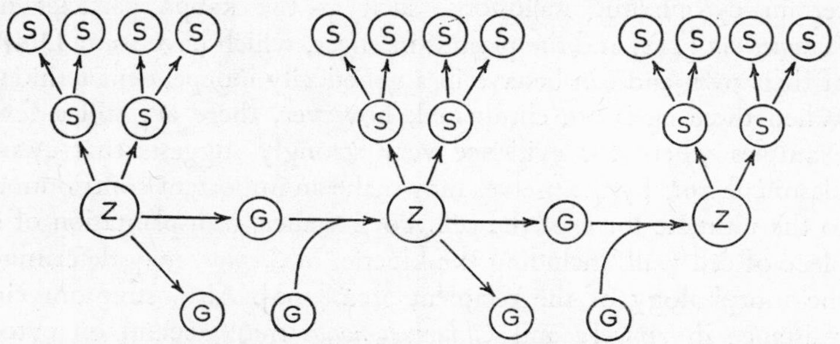

Figure 10.2. Heredity in the cells of multicellular creatures. Continuity from one animal to another is maintained only through the gametes. Somatic cells give rise to other somatic cells in the same individual. (G – gamete; Z – zygote; S – somatic cell.)

through the sex cells (figure 10.2). The highly differentiated somatic cells of each organism are almost invariably developed from the gametes and do not themselves give rise to new organisms. Exceptions are found in some plants and invertebrates, in which an entire creature can develop from a single somatic cell (or a very few cells). The formation of a new plant or animal, or a new organ, from an incomplete fragment, is called regeneration. It is of considerable interest for the whole problem of cytodifferentiation.

Regeneration. Many plants can regenerate from quite small parts. Steward [454] has even claimed to have grown a complete plant from a single carrot cell. Certain plant tissues, particularly meristematic tissues, retain a high degree of developmental flexibility so that they can, as it were, switch on genes which may not have been fully functional in the tissue from which the regenerate arose [328]. A similar property is observed in some of the lower metazoa, such as the *Coelenterata* and planarian worms (*Turbellaria*). In the coelenterates in particular, for instance *Hydra*, an entirely new animal may be produced from a very small piece of the original. In this creature regeneration stems almost entirely from unspecialized cells which are distributed throughout the body. On moving higher up the scale of development in animals the ability to regenerate diminishes progressively.

Much work on regeneration has been done in amphibia since the *Urodela* can regenerate new limbs after amputation. In some other amphibia the ability to regenerate is much less and at the level of the mammals regenerative ability is limited to the capacity to heal superficial wounds and fractures and to replace lost hepatic tissue by compensatory hyperplasia. In these higher forms it seems that regeneration occurs by the multiplication of existing elements whereas, in lower animals and perhaps even in amphibia, highly differentiated cells may revert to a more primitive state before developing into the typical adult cells of regenerated tissue.

Modulation and differentiation. Some of the phenomena described are clearly examples of a completely reversible process, while others represent an apparently different state of affairs. For instance, the mature neuron in higher animals can never, so far as is known, revert to a more primitive form. The term 'modulation' was introduced to describe the reversible process and to distinguish it from 'differentiation' which was assumed to be irreversible. It is questionable whether the two processes can be distinguished so arbitrarily.

The embryological concept of 'potency' is used to describe different degrees of developmental potential, and is more flexible than the idea of 'modulation' versus 'differentiation'. Potency in respect of given characteristics is the capacity of a cell to express the genetic potentialities for these properties. It is one of the maxims of embryology that differentiation involves a progressive restriction of potency. A cell which retains total potency is referred to as totipotent. Each blastomere in an amphibian egg can give rise to a whole individual and is therefore totipotent.

The roles of nucleus and cytoplasm. One of the most important questions in relation to cytodifferentiation concerns the extent to which the nucleus changes spontaneously and independently in the formation of different types of cells, and the extent to which factors in the cytoplasm modify the genetic information carried by the nucleus.

Animal eggs can be divided into two general groups, depending on whether there is already regional differentiation in the single-cell egg or whether differentiation develops only after several cleavages. Eggs in which differentiation of cytoplasm has already occurred even before fertilization are called mosaic eggs. These often exhibit obvious morphological differences in different parts of the cytoplasm and so resemble some Protista, such as *Paramecium* (although the internal structure of the egg is simpler). When cleavage occurs in an egg of this kind, the daughter cells are already different from each other. In insect eggs, for example, the nucleus first divides without any cytoplasmic division. The daughter nuclei then migrate to different parts of the egg before cell walls appear. Damage to a specific region of the cytoplasm of such an egg, even before a nucleus has migrated there, causes a specifically localized lesion in the mature insect. Furthermore, if mosaic eggs are centrifuged so as to redistribute the material within them, a completely chaotic embryo results, with the different tissue layers confused [90]. It is clear, then, that in the mosaic egg the properties of the cytoplasm play an important part in the determination of development. On the other hand, in some species, the egg is of the type known as regulative. In the development of regulative eggs, the first two or four cells produced by cleavage (blastomeres) behave similarly in that a complete and normal, though often small, adult may arise from any one of them if they are separated naturally or by artificial means. That is, the blastomeres are totipotent. In human beings identical twins may arise in this way.

Probably even in regulative eggs there is regional specialization which is important for the early orientation of embryonic development. The regulative egg is thought to be able to readjust (or regulate) a disturbance of this orientation.

With the frog egg, which has some features of both mosaic and regulative eggs, attempts have been made, by micrurgical means, to determine the parts played by nucleus and cytoplasm in differentiation. Curtis [102] transplanted pieces of the cortex from one egg to another intact egg and found that two embryos resulted. One developed in the normal way, the other arose at the area where the transplant was applied. This experiment implies that the egg cortex

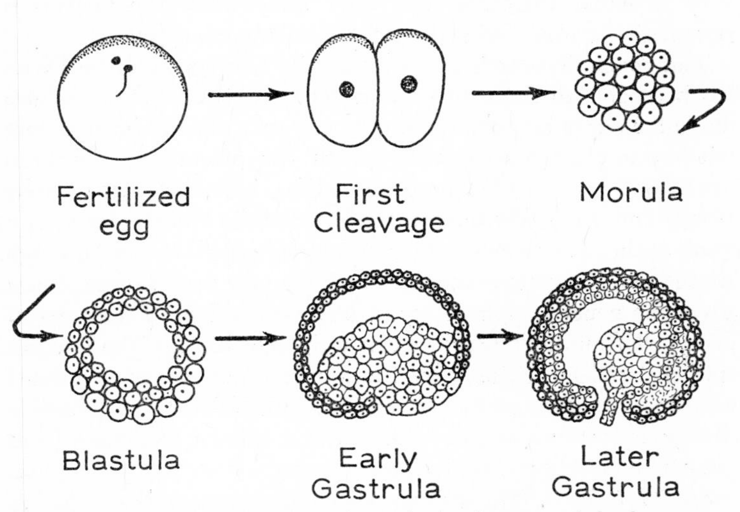

Figure **10.3. The main stages in the early development of the frog egg.**

carries some information for the initiation of embryonic development. Most work of this kind has been directed at elucidating the role of the nucleus. In the classical experiments first performed by Briggs and King [66, 147, 268], nuclei from cells of older embryos were transplanted into unfertilized ova from which the pronucleus had been removed. Their subsequent fate was studied (figures 10.3, 10.4). Nuclei from cells up to the blastula stage were capable of replacing the nuclei of eggs which had been enucleated, and these developed to form healthy tadpoles and frogs. By contrast, if nuclei were taken after gastrulation, development of the egg was usually incomplete and a variety of abortive forms were obtained. Some of the most interesting results came from experiments in which nuclei were transplanted from gastrulae into enucleated eggs which were allowed to form blastulae. Nuclei were then removed from the blastulae and each was put into an enucleated egg. When these eggs were allowed to develop they often formed abnormal embryos, but all the embryos from nuclei from the same blastula were identical and abnormal in the same respect. These experiments suggest very strongly, therefore, that after gastrulation the nucleus is systematically modified and restricted in its potentialities.

This conclusion has been challenged by Gurdon [189, 190] who has been able to show that nuclei removed from the epithelium lining the gut of tadpoles can lead to perfectly normal development when transplanted to ova in which the pronucleus had been inactivated by radiation. In many species, including that used by Briggs and King, chromosomal abnormalities readily occur as a result of the operative procedure and are inherited. Hence, although Briggs' and King's experiments paved the way there is some doubt about the general applicability of their conclusions. The nuclei of many cells from adult tissues may be totipotent, like the Xenopus gut epithelial cell nuclei, used by Gurdon; it is not clearly established whether the failures are due to technical difficulties or to irreversible changes in the nucleus. In any event it is of interest that nuclei from gut epithelial cells remain totipotent.

Specific inducers. The stimuli which initiate cytodifferentiation have commanded a good deal of attention. It has been amply demonstrated that, as soon as different kinds of cells have appeared in the embryo, they begin to interact with each other in a specific manner. For example, many ectodermal structures develop normally, only if the cells are in contact with the appropriate mesodermal or endodermal tissue [300, 424]. In Spemann's [443] classical experiments it was shown that teeth developed only from ectoderm in contact with mouth endoderm. Ectoderm from any part of the animal could be induced to form teeth if transplanted on to mouth endoderm at an early stage in development. After induction, however, the ectoderm became 'fixed' as tooth-producing ectoderm.

Many systems of this kind have been analysed and the very high specificity of the interactions has been conclusively proved.

The chemical nature of inducers and 'organizers' has been the

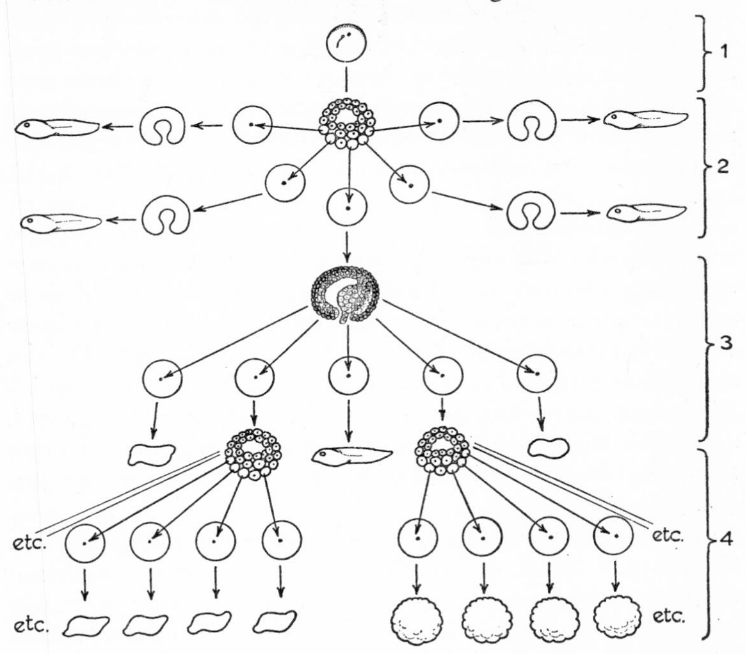

Figure 10.4. Briggs' and King's experiments on nuclear transplantation in amphibian embryos. 1. A fertilized egg is allowed to grow to a blastula. 2. Nuclei from blastula cells are transplanted to unfertilized eggs from which the pronucleus has been removed. These eggs develop and form identical tadpoles. 3. If eggs are allowed to develop beyond early gastrula before transplanting nuclei from them to enucleated eggs, the resulting embryos do not develop normally but are arrested at different stages. 4. If nuclei are removed from these embryos at the blastula stage and transplanted to enucleated eggs then all the embryos derived from one blastula are arrested at the same stage, demonstrating that some capacity of the nucleus has been lost during development to late gastrula.

subject of much investigation and speculation. The field has, unfortunately, been very much confused by the discovery that many specific interactions can be simulated by treating embryonic tissues with non-specific substances. Despite this confusion two classes of substance – ribonucleic acids and proteins – have been repeatedly implicated as specific inducers and it is possible that both may play a part.

The evidence for ribonucleic acid is derived mainly from experiments which showed that ribonuclease would interfere with development [54]. Interest in these observations has recently been renewed as a result of experiments which have demonstrated that, when foreign RNA is added to culture medium, cells may produce proteins characteristic of the source of the RNA [349, 350, 351].

Strong experimental evidence has also accumulated for proteins as specific inducers in certain systems. For example, cartilage formation from presumptive cartilage cells can be induced in tissue culture on the application of extracts from notochord. These extracts have been highly purified, and an electrophoretically and chromatographically pure protein fraction has been obtained which will induce differentiation in minute amounts [227, 228, 277, 278]. Substances which induce the development of mesodermal structures have also been purified and shown to be proteins [28, 29, 473, 475, 517].

In at least one system a fat-soluble vitamin acts as an inducer. Vitamin A, added to medium in which chick embryonic ectoderm is grown, stimulates the formation of ciliated, mucus-secreting epithelium [138, 139]. In its absence keratinized epithelium forms. Hormones can also act as inducers [515].

These results indicate that embryonic induction is not a property of a restricted class of substances. They suggest that differentiation may be the result of complex mechanisms which can be influenced at different points.

Possible mechanisms of cytodifferentiation

It seems rational to try to explain the phenomena of differentiation in terms of regulatory mechanisms already known in cells. At the outset it may be pointed out that all the phenomena of cytodifferentiation can be explained by stable feedbacks such as are well recognized in electronic and other engineering control systems

[168]. To take a simple example, if the end products of one reaction inhibit another reaction whose end products can, in turn, inhibit the first one then there are two possibilities: (i) the two reactions may exist simultaneously in a precarious equilibrium; or (ii) one may exclude the other. In this simple way a kind of 'flip-flop' arrangement can exist. More complicated systems are easily envisaged whereby several complexes of reactions may be mutually exclusive. It is therefore possible to construct a model cell in which differentiation results from temporary stimuli at some stage during development, and which may be completely irreversible and independent of the genetic material in the cell. Controls of this kind may operate at any functional level; the possibilities will be discussed in relation to the chemical control mechanisms described in chapters 6 and 7.

(i) The first possibility is that cytodifferentiation results from a progressive loss of genetic material. There is no doubt that this can occur. For instance, in the gall midge *Mayetiola destructor*, most of the chromosomes are lost during somatic differentiation, and indeed only the sex cells in the adult have a full complement. For some time it was suspected that this might be a general mechanism of cytodifferentiation but this theory is now out of favour since in most organisms the number of chromosomes and the amount of DNA present in somatic cell nuclei is exactly the same as in the newly fertilized egg [44, 482]. However, the fact that chromosome deletion does occur in a few species indicates that cytodifferentiation may result from more than one mechanism.

(ii) The second possibility is that during cytodifferentiation the capacity of DNA to act as a template for RNA is progressively restricted. In chapter 7 it was indicated that the most likely explanation for enzyme induction is that the functioning of certain genes is inhibited by substances (regulators or repressors) whose nature was at that time not specified but which prevented DNA from acting as a template for RNA. Most of the phenomena of cytodifferentiation could be explained by postulating that many genes are inactive, owing to 'masking' by some substance, possibly similar to the regulator substance involved in enzyme induction.

There is some direct evidence for this. For instance, in the giant salivary gland chromosomes of certain insects (chapter 8) periodic swellings occur. These swellings are found in specific regions of certain chromosomes at specific stages of development (plate 11C, D) [30, 31]. They are known as 'puffs', or 'Balbiani rings' and are considered to reflect the functioning of parts of the chromosomes. By auto-radiography it has been shown that there is very active RNA synthesis in some of these regions, whereas RNA synthesis in other parts of the chromosome may be very slow [143]. Moreover, if chromosomes are dissected from the gland of a prepupal stage and put into egg-contents from a preblastoderm stage one of the normal 'puffs' subsides while a puff appears on a different part of the chromosome [275, 276]. This suggests that RNA synthesis in puffs is subject to specific regulation by substances in the cytoplasm. In the lampbrush chromosomes of amphibian oocytes it has also been found that RNA synthesis goes on steadily in certain regions although it is hardly detectable in others.

Support for the masking theory has been obtained by studies on chromatin isolated from rabbit thymus gland and bone marrow nuclei. Using 'molecular hybridization' techniques Paul and Gilmour have shown that only 5–10% of all the different DNA molecules in this material are available, as templates for RNA synthesis; the remainder seem to be masked. Moreover, different sets of DNA molecules are available in different organs; hence, the masking is organ-specific [366, 367].

In seeking a substance which might be responsible for the blocking of mRNA synthesis most interest recently has centred on histone [8]. The histones are present in very constant amounts in somatic cells. Furthermore, histones inhibit the synthesis of RNA on a DNA template and their removal from chromatin increases its effectiveness as a template for RNA synthesis. Studies of chromosomally directed protein synthesis in cell-free systems also point to the general conclusion that masking of DNA is mediated through histones. How specificity is achieved is not known.

(iii) A third possibility is that mRNA is formed by all the DNA in a chromosome but is very rapidly broken down again unless adequately stabilized. It has been suggested that histones perform the function of stabilizing mRNA [290], and that unprotected

RNA is almost immediately destroyed by ribonuclease whereas protected RNA goes on to perform its function.

(iv) A fourth suggestion is that the control may operate at the level of the ribosome [518]. According to this hypothesis the newly formed protein molecule remains bound to the mRNA-ribosome complex, unless an inducer is present – in which case it is liberated. The formation of new protein would then be limited because the templates are already occupied. A refinement of this theory suggests that the formation of mRNA is itself regulated by some kind of feedback and that a new mRNA molecule is formed only when another one of the same type has been broken down.

(v) Finally, many of the phenomena of cytodifferentiation could be explained simply by stable feedback systems affecting either the enzyme systems themselves or enzyme-forming systems.

The evidence for the masking theory is now good but this does not exclude the possibility that some or all of these other mechanisms may also operate in cytodifferentiation.

11. Cellular Interaction

In the development of a multicellular organism from a fertilized germ cell the end result is a colony of variable size consisting of large numbers of cells, often of many different types, performing very different functions. These operate together in a harmonious fashion, and hence there is a very large measure of functional interdependence.

Special mechanisms are required to ensure that cells of similar types remain together and associate with cells of other types in the correct spatial relationships. This applies, not only in the course of embryonic development, but also in adult life: it is easy to show by removing part of an organ that during repair the cells can exhibit considerable motility. Also, when tissues are removed and cultured, motility of the cells in the explant is often a striking feature. Hence most cells in an adult organism are maintained in relation to each other as a result of continuously operating restraints to their normal tendency to migrate. This is perhaps most dramatically demonstrated by an experiment in which *Hydra* is made to turn inside out so that the endodermal tissues are on the outside and the ectodermal tissues on the inside. The normal relationships are restored by active migration inwards of endoderm and outwards of ectoderm [413].

Cell contact and adhesion

In the formation of a multicellular organism the key process is the aggregation of individual cells to form colonies. To aggregate, cells must first come into contact, either with each other or with a common matrix, and they must then adhere [198]. The process of adhesion involves a measure of specificity since like cells tend to adhere to each other. The molecular basis of this specificity is not yet understood, but the result is the orderly arrangement of cells to form tissues in the developing embryo. There is a good deal of evidence to suggest that one of the most characteristic features of cancer cells is a loss of this property of specific adhesion [2].

Cells are maintained in relation to each other as a result of at least three recognizable types of contacts:

(i) inclusion in a common matrix or adhesion to a common matrix;
(ii) contact without any intercellular matrix material;
(iii) direct protoplasmic connections between cells.

(i) The intercellular matrix in most tissues consists of such substances as hyaluronic acid, chondroitin sulphuric acid and cellulose which are characteristic of cell walls and have already been discussed under that heading (chapter 3). It is not certain whether these materials are themselves responsible for the specific pattern of aggregation, or whether they merely accumulate between cells of certain types after the initial aggregation has taken place [38]. Most of the material certainly accumulates after cells have made contact with each other; but the important question is whether aggregation precedes or follows their first appearance.

(ii) In many cellular tissues there appears to be no intercellular material, or only a very small amount (which may have accumulated since the cells came into contact). On electron microscopic examination, it is found that the gap between the outer regions of the cell membranes of two adjacent cells in these tissues is always of the order of 150 Å. Where it is different from this it is usually 20 Å–50 Å (the so-called 'close gap', which is much less common).

These dimensions are extremely constant and have led investigators to consider the possibilities of such cell contacts being due to physical properties of the membrane itself [103, 104, 212, 384]. Indeed, these gaps can be explained by the fact that the attractive and repulsive forces of cell membranes would be expected to form stable gaps at about these distances, and oil-water models have been found to simulate the gaps almost exactly [481].

Many of the known properties of cell membranes fit in with such a mechanism. Most cells have a nett surface negative charge, which can be measured electrophoretically. Cells which have a particularly high surface charge (such as erythrocytes) do not normally aggregate. Some other cells which have a rather high surface charge and do not aggregate easily can be made to do so by treating them with a positively charged polyelectrolyte, such as polylysine [342]. It is suggested that the role of bivalent cations (such as calcium) in cell adhesion is a result of the same effect. On removal of bivalent cations by chelating agents (such as ethylenediaminetetra-acetic acid) many cells can be caused to separate from each other: for instance, kidney and liver tissue can be made to disaggregate by this means alone. (The suggestion occasionally made, that calcium forms bridges between cells, cannot possibly be true, since the intermembrane distances would then be of the order of only a few (2–3) Å instead of about 150 Å.)

The role of enzymes in disaggregating cell masses is more difficult to explain on this basis. Some enzymes, such as trypsin, are effective as chelating agents, and this was at one time thought to be their mode of action. However, trypsin is still effective in the presence of high concentrations of bivalent cations. It was also suggested that proteins might act simply by being adsorbed to the cell membrane. However, while this may be true during cell adhesion (as discussed below), it is probably not an adequate explanation for the action of proteolytic enzymes, since they are no longer effective in disaggregating cells when the active groupings are blocked by diisopropyl fluorophosphate [122]. The enzymes which are most effective in separating cells are proteases and mucases; this reinforces the evidence, derived from other studies, that mucoproteins are involved in cell contacts.

The question arises whether it is possible to combine the two theories. Simple physical forces may account for the spacing between cells while a substance, possibly mucopolysaccharide, may account for the specificity of cell association. Curtis [104] points out that the interpolation of protein layers would weaken the attraction forces and he therefore does not favour this suggestion. On the other hand, there is much suggestive evidence from other work, including the electron microscopic data already referred to which would favour this idea. Furthermore, the figures available for electrostatic and van der Waals forces in cell membranes are very approximate, so that the combined theory might fit most cases. It has many appealing features and is the one most generally accepted at present.

When cells remain in close apposition (with 150 Å gaps) for some time, they develop areas of increased density opposite each other in the sub-membranous cytoplasm. These are the structures called *desmosomes*. Their nature and functions are obscure but they do not represent direct cellular contacts as was at one time thought [131].

(iii) Direct cellular connections are clearly more common than was thought until very recently. In plants cytoplasmic bridges or plasmadesmata between cells have been recognized for a long time. These provide very narrow connections, of the order of 100 Å, between cells. Ions and small particles can probably pass readily through these but particles of the size of ribosomes probably cannot.

Quite recently it has been shown that in many animal tissues there may be protoplasmic connections of a similar nature since free movement of ions throughout the tissues can be detected [295, 377, 373]. These may be of importance in ensuring the harmonious functioning of the cells in a tissue.

In many cases the connections are much more extensive and give rise to a true syncitium as in the skeletal and cardiac muscle of many animals.

Cell aggregation

When a sponge is pushed through fine gauze the cells are separated and may continue to survive as individuals for some time. If they are allowed to mix together, they stick to each other and aggregate to form a new sponge [447]. Holtfreter showed that amphibian embryonic material could be disintegrated in a similar manner by placing it in a calcium-free medium. When the separated cells were returned to a medium containing calcium they reaggregated and reconstituted the original embryonic structure [225, 226]. Experiments of this kind have been carried further, especially by

Moscona [186, 329, 330, 331, 332, 333] with tissues of higher organisms. He has shown that a piece of embryonic tissue, for example kidney, may be disintegrated by treating it with trypsin until it forms a homogeneous suspension of individual cells. When these cells are gently shaken together they first aggregate to form a mass in which the different kinds of cells are dispersed randomly. Soon, however (24–48 hours), the cells segregate into groups of similar type. For instance, mesodermal cells characteristically migrate to the centre of the mass and ectodermal ones to the outside. In reaggregates of kidney cells typical kidney tubular structures may ultimately be formed. These experiments indicate that individual cells have the ability to 'recognize' each other.

The recognition is tissue-specific rather than species-specific, provided the species are not too distant. For instance, chick and mouse tissues will form chimaeric reaggregates in which cells of each type (recognizable by their nuclear morphology) segregate together irrespective of species. On the other hand, this is a limited generalization since disaggregated cells from two different species of sponge will segregate into separate reaggregates according to species; chimaeric aggregates never form.

An essential component of a reaggregating animal cell system appears to be an extracellular material (ECM) which is probably of mucopolysaccharide nature. Moscona [333, 334] has convincingly shown that reaggregation depends on the presence of this substance, although it has also been suggested that self-recognition during aggregation is an inevitable consequence of non-specific physical characteristics of the cell surface [451].

Recognition of like and unlike is probably due to a surface property of the cells, the nature of which is still not understood. The phenomenon has to be distinguished from the immunological self-recognition reactions which develop later and may or may not be closely related [69].

Contact inhibition, as described by Abercrombie and Heaysman [1, 3, 4], may be related to the behaviour of aggregating cells. These investigators studied normal fibroblasts by time-lapse cinemicrography and observed that when cells came into contact with one another there was an immediate reaction expressed by a sudden immobility of the cytoplasm in the region of contact. This phenomenon was demonstrable only in normal fibroblasts and could not be shown in sarcoma cells. It is considered to play an important part in the development of the parallel alignment of cells which is typical of normal fibroblastic tissues.

Inductive interaction

It was discovered many years ago that, during embryological development, some parts of the embryo induced specific structures in other parts. This inductive interaction is of at least two types. One is associated with the initiation of cytodifferentiation, as described in the previous chapter. The other is involved in morphogenesis or tissue organization.

If mesodermal and ectodermal elements of a tissue are carefully separated and cultured independently, they form undifferentiated sheets; whereas, when they are grown together, the ectodermal cells form tissue-specific structures such as glandular tubes and acini [19, 182, 183, 184, 302]. Grobstein, in particular, has made a special study of this and has shown that, while induction cannot be obtained through a cellophane membrane, it can be obtained through a millipore membrane (which has large numbers of minute channels about 0·25 μ in diameter).

In these experiments, ectoderm was separated from mesoderm by gentle tryptic digestion and the two components of the tissue were explanted on opposite sides of a membrane. Submaxillary gland mesoderm caused ectoderm of the same gland to develop acinar structures. Similarly, when kidney mesoderm and kidney ectoderm were grown on opposite sides of a membrane, the ectoderm was induced to form tubular structures. On the other hand, when kidney mesoderm was grown with submaxillary gland ectoderm, no inductive effect was obtained. These experiments show that the inducing effect is largely tissue-specific. The material which passes through the millipore membrane has not been positively identified, but is probably a mucopolysaccharide.

Other morphogenetic interactions

Closely related to aggregation is the phenomenon of morphogenetic migration. During embryonic development some cells migrate from the area in which they originally appear and move through the embryo until they come to specific regions where they settle down [501, 502, 503, 504]. It is in this way that, for instance, the cells of the nervous system are distributed throughout the body. Again some specific mechanism for the recognition of specific kinds of cells must operate. As a further development of migration the folding movements which occur within the embryo are, of course, of the greatest importance in giving rise to its final form.

Four factors have been recognized as operating in morphogenetic migration. Two of these, contact inhibition and aggregation, have already been described; they restrict cell movement. The other two, chemotaxis and contact guidance, promote and guide cell movement. Chemotaxis is the attraction exerted upon cells by chemical substances. This is exemplified by the movement of macrophages towards bacteria. Contact guidance is the movement of cells along linear irregularities, such as stretched fibrin threads or lines scored with a diamond on glass. Other unknown factors may also play a part in morphogenetic migration and much has still to be learnt about the phenomenon.

Finally, among morphogenetic interactions must be mentioned cell death. In constructing the architecture of tissues it is not uncommon for one cell type to form an initial scaffolding which later breaks down as mature cells move in. This phenomenon is particularly marked in metamorphosis, for instance in tail resorption in Amphibia, but it occurs during embryonic development in almost all animals.

Homeostasis in the adult

As pointed out earlier, the mechanisms which control many of these reactions must persist into adult life and all tissues can be regarded as being in a state of dynamic equilibrium. Cells are dying and being replaced continuously in most tissues and, in some, cells are migrating in and out all the time.

Not only is migration controlled by local interactions but cell growth is similarly regulated. Many tissues with a low rate of cell multiplication begin to grow very rapidly when continuity is severed. An outstanding example is the liver, which normally contains very few dividing cells. If a large part of the organ is removed by operation there is an almost immediate shift to a high rate of cell division and the cell mass may double in a little over 24 hours. No useful information has as yet been obtained concerning the nature of the control processes in cell division, but they probably involve metabolic feedbacks. It could be argued, for instance, that substances involved in feedback inhibition of nucleic acid synthesis might accumulate to inhibitory levels in intact tissues but leak out if tissue continuity were severed, so removing the inhibition. Mechanisms of this kind inhibit growth in some bacterial cultures, but no convincing evidence for similar mechanisms in animal and plant cells has been discovered.

In a complex organism most cells are functionally dependent on others. Regulation of the metabolism of all tissues so that they operate harmoniously requires special mechanisms besides the local interactions which have been described. These mechanisms are of two types. First, there are those which operate by hormonal regulation and metabolic feedbacks and control the overall metabolism of the whole organism (for example, the pituitary–adrenal system in Vertebrates). They are quite well understood. There must also be others, however, which maintain the balance among different organs and are revealed, for example, by compensatory hypertrophy of the remaining organ when one of two paired organs (such as the kidneys) is removed. Although little is known about these there is growing evidence that they are mediated by hormones also. Plant hormones (auxins) have clearly been identified as performing this role. For example, indole acetic acid promotes root growth and kinetin promotes growth of shoots in appropriate conditions. In insects, hormones regulate metamorphosis; the juvenile hormone controls maturation and the pupating hormone, ecdysone, is involved in the control of pupation. In amphibia, too, the thyroid hormone, which regulates the basal rate of energy metabolism in cells, is needed for metamorphosis. The most thoroughly investigated tissue-specific hormone is erythropoietin which is produced by special cells in the kidneys of mammals in response to low oxygen tension. This hormone induces maturation of erythropoietic stem cells into mature red blood cells. Some recent evidence suggests that liver size may be similarly controlled by a hormone.

There is still very little information about the mode of action of hormones on cells. Some, such as insulin, seem to act on the cell surface. Ecdysone probably does too. It has been suggested that steroid hormones may also act at this site by replacing cholesterol in unit membranes [512]. However, this attractive idea has apparently been disproved [165]. There is evidence that some steroid hormones act directly on enzyme systems [123, 483] but evidence has also been adduced recently to suggest that they may have an effect on protein synthesis, through the genetic mechanism [180]. The effect of thyroid hormone is also prevented by substances which stop the synthesis of messenger RNA [463]. However, since it is known that thyroid hormone affects mitochondria directly it is difficult to understand the significance of this observation.

Part Five: The Origin and Evolution of Cells

12. The Origin and Evolution of Cells

Origin of organic substances

There has been some recent speculation about the way in which cells arose. It is doubtful whether we will ever be certain about the actual course of events but a plausible picture has now emerged which receives a good deal of support from established facts [74, 148, 356, 357, 489].

When the interstellar gases and dusts condensed around the sun and this planet was newly formed it had an atmosphere which consisted almost entirely of hydrogen and methane [479]. Because of the relatively low gravitational field of earth and its relatively high temperature much of the hydrogen was soon lost in space. Nevertheless, the early atmosphere was probably still strongly reducing, and contained such gases as methane, carbon dioxide, hydrogen, nitrogen, water, hydrogen sulphide and ammonia.

The hydrosphere (the watery layer surrounding the earth) probably consisted mainly of an aqueous solution of these substances with some ions, such as phosphate and chloride, washed out of the rocks.

Because there was little oxygen in the atmosphere there could have been no ozone layer such as now exists at the outer fringe of our atmosphere; this is produced by the ionizing effect of ultraviolet light on oxygen. The ozone layer effectively filters out short ultraviolet light rays from the sun; therefore, these must have penetrated the atmosphere of the earth much more deeply in the beginning. Conditions were consequently created for the abiogenic synthesis of many of the molecules we associate with living processes, such as carbohydrates, amino acids and fatty acids. Most of these molecules are predominantly partly reduced compounds of carbon or nitrogen. It was shown in 1955 by Miller [318] that when an electric discharge is passed through a mixture of methane, ammonia, water and carbon dioxide, some carbohydrates, amino acids and fatty acids are formed. Other investigators have demonstrated that the same result can be obtained by irradiating a similar mixture with short ultraviolet light [372]. Consequently, it seems reasonable to postulate that, at a very early period, probably 2×10^9 years ago, there was a fairly substantial syntheses of molecules of this type in the absence of true biological reactions.

Subsequent events are more speculative but there is some experimental evidence to suggest various possibilities. For instance, if amino acids are allowed to absorb on to certain clays they polymerize and form polypeptides. It can therefore be postulated that protein-like substances first arose in this way.

Oparin estimates that the concentration of abiogenically synthesized organic substances in the 'primeval broth' was quite high. In any event, for life to evolve further it was clearly necessary that these protein, polypeptide, carbohydrates and lipid substances should be concentrated together. There are two reasonable possibilities to account for this. One theory suggests that monolayers of lipid stretched over parts of the primitive sea and absorbed protein to form lipo-protein monolayers. By the action of the wind small droplets could easily be formed from these giving rise to lipid-enclosed droplets containing protein. Alternatively, lipids may not have been necessary in the early stages of development because De Jong has demonstrated that dilute solutions, containing mixtures of certain proteins and carbohydrates, lead to the formation of coacervates [48]. Coacervates are formed by molecules which have a strong tendency to coalesce so that most of the material initially in solution becomes concentrated in small droplets. These droplets may be quite complex, and some, which have been produced experimentally, may include structures very similar to those found in cells. Possibly both mechanisms contributed to the formation of the so-called eobionts, which preceded the first real cells.

There is a big hiatus between the formation of organic coacervates and the appearance of the first true living cells. Proteins capable of performing specific catalytic functions (that is, enzymes) had to evolve either by accident or by some kind of selective process. Furthermore many functionally related enzymes had to be concentrated in a single coacervate in order to catalyse even the simplest reactions. A system of replication, involving the nucleic acids, must also have developed. No convincing theory concerning the appearance of the nucleic acids has yet been proposed but it is extremely difficult to postulate any reasonable evolutionary pattern for proteins in the absence of nucleic acids. It therefore seems likely that, at a rather early date in evolution, the association between nucleic acid and protein structure became established. Possibly, in the early stages, the polymerization of amino acids on a nucleic acid template occurred by random collision and was not catalysed in any way. If so, evolution must have proceeded very slowly indeed at this stage. However, since the period from the formation of the earth to the appearance of cells was perhaps 10^9 years there was plenty of time.

Protobacteria

By the end of this era, almost 10^9 years ago, it is virtually certain that life was effectively established, because it was at this time that oxygen first appeared in relatively large amounts in the atmosphere of the earth, and there is every reason to believe that nearly all the oxygen in the earth's atmosphere has been produced by photosynthesis. Before the development of photosynthesis, primitive organisms, sometimes referred to as Protobacteria, must already have arisen and perhaps even reached a relatively high level of complexity. It seems likely, for instance, that they had already developed a reproductive system, involving nucleic acids, and that the relationship between nucleic acid and protein structure was well established. These organisms probably had a glycolytic type of metabolism [121]. Moreover, since most of the amino acids could have been synthesized abiogenically in the primeval broth it is virtually certain that the protobacteria were heterotrophs (dependent on the availability of ready-made organic substances). Metabolic pathways utilizing ATP, carbamyl phosphate, nicotinamide nucleotides and similar molecules were probably well established. Most of the metabolic processes which had evolved were probably based on dehydrogenations; consequently sulphur, iron, nitrogen and a few other substances probably acted as hydrogen acceptors in the absence of oxygen. Carbon dioxide fixation was probably already established since it is one of the most widespread properties of living matter. Its close association with photosynthesis probably arose at a later date.

Accumulation of oxygen. Before photosynthesis began there is no doubt that some oxygen was produced, either by photolysis of water in the upper atmosphere or by release of bound oxygen from some of the rocks of the lithosphere. Oxygen released from rock was almost certainly reduced very quickly by iron, but some of the oxygen formed by photolysis may have accumulated, since free hydrogen formed at very high altitudes would escape into space. However, by far the greatest production of free oxygen probably followed the appearance of photosynthetic pigments which permitted photolysis of water by visible light. As oxygen accumulated in the atmosphere an ozone layer must have become established at its outer surface, and this would be expected to limit those reactions which were catalysed by ultraviolet light, since the short ultraviolet radiations would be absorbed by it, as they are at the present time.

Oxygen inevitably became an important hydrogen acceptor and consequently the aerobic pathways of metabolism, giving rise to high energy phosphate, were established. At about this time the abiogenic synthesis of amino acids and similar compounds must have ceased, owing to the absorption of ultraviolet light in the ozone layer. Consequently biogenic syntheses of these reduced compounds of carbon and nitrogen must have evolved, conferring considerable survival value on the organisms possessing them. It was possibly about this time therefore that the autotrophic organisms began to appear (organisms capable of utilizing inorganic substances). Subsequently, it seems very likely that organisms possessing complementary metabolic functions became associated symbiotically and that eventually, by fusion, these gave rise to more highly organized cells.

Much of this account of the origin of simple organisms is, of course, highly speculative but with the emergence of the bacteria we can propose a more convincing and realistic kind of evolution.

Specialization

So far we have been considering only the acquisition of new properties by simple organisms. By this process alone there must have arisen a great variety of forms. One consequence must have been that certain organisms were, from the beginning, nutritionally dependent on others. For instance, with the appearance of oxygen in the atmosphere and the termination of abiogenic syntheses, nonphotosynthetic organisms must have become entirely dependent on those which could utilize light energy. Thus, as the environment changed, the acquisition of new properties by certain organisms was essential for the survival of all other living forms and this itself must have initiated many interdependent relationships.

From the very beginning however an opposite process must have been going on, that is, organisms must have been losing properties. We can say this with certainty because we can observe it happening continuously in experimental conditions. 'Loss mutations' are indeed very much more frequent than 'gain mutations' in our present biosphere. Not all are necessarily harmful. For instance, if organisms lose the ability to synthesize a substance which is available in the environment then they suffer no harm; indeed their metabolism might be considered to function more economically.

The great majority of mutations may not fall into the categories of 'gain' or 'loss' mutations at all. We now know that many mutations involve substitution of one amino acid by another as a result of a change in a single nucleotide; the altered protein may be greatly, only slightly, or not at all different in function from the old. (See the discussion of Yanofsky's work in chapter 8.) Mutations of this kind permit a slow drift of structure during evolution. Most proteins may exist in structurally different but functionally identical forms in different species. (Heteroenzymes are examples of this.) In particular, the structure of the cytochromes has been studied in many organisms [306] and similarities have been found even between cytochrome Cs from yeast and human material. The more closely organisms are related (i.e. the more recently they have diverged from a common ancestor in the course of evolution) the fewer are the differences observed. Conversely, the more distant the relationship the greater the differences. For example, the horse and pig, which are thought to have diverged about 33 million years ago, have three differences in their cytochrome C, the horse and chicken have twelve differences, while the horse and yeast, which probably diverged 500 million years ago, have no fewer than forty-four differences, although recognizable similarities remain.

Besides these differences between species structurally different but functionally similar proteins may persist in the same organ. Isozymes are examples. Some isozymes are indistinguishable in function but others exhibit clearcut differences which often endow the two proteins with unique properties although they have similar general functions. (See, for example, the discussion of aspartokinase isozymes in *E. coli* in chapter 7.) It is not difficult to imagine how the several cytochromes in the electron transport system (chapter 4) could have evolved from a common ancestor, or even how myoglobin and haemoglobin (chapter 2) could have arisen by divergence from a common progenitor in the course of evolution.

Hence, the acquisition of new properties and the loss of others probably proceeded simultaneously even from the earliest stage of evolution and, at a later stage, particularly in the formation of the so-called higher forms, a kind of specialization probably developed whereby certain properties were given up in order to permit those that remained to function more efficiently [296]. Even among bacteria some interesting examples of this kind of specialization have been demonstrated. For example, table 4 shows the nutritional relationships in some of the *Mycobacteria*, among which an increasing trend towards parasitism is associated with a progressive loss of the capacity to use different carbohydrates as a sole carbon source.

As a consequence both of the acquisition of new properties and the loss of old ones, therefore, symbiotic relationships of great complexity have developed. Examples of typical simple symbiotic relationships are shown in table 5.

Table 4. Carbon source utilization by some *Mycobacteria* (from Lwoff [296]), showing the trend towards monotrophy (dependence on a single source) which is correlated with the development of parasitism.

Carbon source	*Saprophytic*		*Parasitic*				
			Cold-blooded		*Warm-blooded*		
	M. phlei	*M. smegmatis*	*M. ranae*	*M. cheionei*	*M. avium*	*M. bovis*	*M. hominis* H37
Glycerol	+	+	+	+	+	+	+
Glucose	+	+	+	+	+	+	±
Pyruvate	+	+	+	+	+0	+	0
Lactate	+	+	+	+	0	0	0
Succinate	+		+	+	0	0	0
Malate	+		+	+	0	0	0
Citrate	+	+	+0	+	0	0	0
Formate	+	+0	0	0	0	0	0
Acetate	+	+	+	+	±	±	0
Propionate	+	+	+	+	0	0	0
Oxalate	+	+0	+	+	0	0	0
Ethanol	+	+	+	0	0	0	0
Fructose	+	+	+	0	0	0	0
Mannitol	+	+	+		0	0	0
Arabinose	+	+	±	0	0	0	0
Galactose	+	+	±	0	0	0	0

Polytrophy ⟶ oligotrophy ⟶ monotrophy

Table 5. Examples of symbiosis among *Protista* (Lwoff [296])

Organisms	*Growth factors required*	*Substances synthesized*	*Growth in medium deficient in growth factors*
Haemophilus parainfluenzae	NAD and NADP	Haematin	0
Haemophilus canis	Haematin	NAD and NADP	0
Haemophilus parainfluenzae and Haemophilus canis			+
Polyporus adustus	Thiamine	Biotin	0
Neumatospora gossypii	Biotin	Thiamine	0
Polyporus adustus and Neumatospora gossypii			+
Rhodotorula rubra	Pyrimidine	Thiazole	0
Mucor ramannianus	Thiazole	Pyrimidine	0
Rhodotorula rubra and Mucor ramannianus			+

The more complicated cells of higher organisms may have originated first from a symbiotic relationship between different organisms which ultimately led to their sharing the same cytoplasm. The most convincing piece of evidence in support of this idea concerns the behaviour of plastids (chloroplasts and related structures) in plant cells. These can behave in a genetically independent manner. There is some reason to believe that mitochondria may exhibit similar characteristics. Certain mutants of yeasts and *Neurospora* exhibit a defect of the mitochondria and this behaves as if genetically independent of the cell nucleus. Finally, DNA has now been isolated from both chloroplasts and mitochondria [338, 339, 415].

Hence it is possible that the single-celled Protozoa and Protophyta already represent a level of interdependent organization involving symbiotic protobacteria (see plate 18).

The interdependence of cells in multicellular organisms is particularly interesting because their specialization represents a sacrifice of certain general functions in order to develop some specialized ones. The great difference between specialized cells of multicellular organisms and the simple single-celled organisms is that, whereas in the latter the loss of a function usually represents a genetic loss which is unlikely ever to be restored, in the cells of multicellular organisms the full genetic potential may be present but only partly realized.

The evolution of the cells in the biosphere has, then, involved a continuous simultaneous gain and loss of properties. In consequence there has been a great shuffling and assortment to yield the enormous variety of cells in existence.

Although this has inevitably resulted in a very great measure of divergence, one of the most striking facts to emerge from the study of cells in recent years is that the same common properties can be identified in almost all of them.

Bibliography

General texts on Cell Biology

ANFINSEN, C. B., 1959. *The molecular basis of evolution:* John Wiley, New York.

BALDWIN, E., 1957. *Dynamic aspects of biochemistry*, 3rd ed.: Cambridge University Press, London.

BRACHET, J., 1957. *Biochemical cytology:* Academic Press, New York.

BRACHET, J., and MIRSKY, A. E. (Eds.), 1959–61. *The Cell*, vols. 1–5: Academic Press, New York.

DAVIDSON, J. N., 1960. *The biochemistry of the nucleic acids*, 4th edn.; Methuen, London.

DE ROBERTIS, E. D. P., NOWINSKI, W. W., and SAEZ, F. A., 1960. *General cytology*, 3rd edn.: Saunders, Philadelphia.

ENGSTRÖM, A., and FINEAN, J. G., 1958. *Biological ultrastructure:* Academic Press, New York.

FINEAN, J. G., 1961. *Chemical ultrastructure in living tissues:* Thomas, Springfield, Illinois.

GIESE, A. C., 1957. *Cell physiology*: Saunders, Philadelphia.

KUYPER, CH. M. A., 1962. *The organization of cellular activity:* Elsevier, Amsterdam.

MCELROY, W. D., and GLASS, B. (Eds.), 1957. *The chemical basis of heredity:* Johns Hopkins Press, Baltimore.

MCELROY, W. D., and GLASS, B. (Eds.), 1958. *The chemical basis of development:* Johns Hopkins Press, Baltimore.

PERUTZ, M. F., 1962. *Proteins and nucleic acids:* Elsevier, Amsterdam.

PICKEN, L., 1960. *The organization of cells and other organisms:* Clarendon Press, Oxford.

RUDNICK, D. (Ed.), 1955. *Developmental cytology:* Ronald Press, New York.

RUDNICK, D. (Ed.), 1956. *Cellular mechanisms in differentiation and growth:* Princeton University Press, Princeton, New Jersey.

SRB, A. M., and OWEN, R. D., 1957. *General genetics:* Freeman, San Francisco.

SWANSON, C. P., 1957. *Cytology and cytogenetics:* Prentice Hall, Inc., Englewood Cliffs, New Jersey.

Bibliography

WADDINGTON, C. H., 1959. *Biological organization: cellular and subcellular:* Pergamon Press, London.

WALKER, P. M. B. (Ed.), 1960. *New approaches in cell biology:* Academic Press, London.

ZIRKLE, R. E. (Ed.), 1959. *A symposium on molecular biology:* Chicago University Press, Chicago.

References

1. ABERCROMBIE, M., 1961. 'The bases of the locomotory behaviour of fibroblasts': *Exper. Cell Res.* Suppl. 8, 188–98.
2. ABERCROMBIE, M., and AMBROSE, E. J., 1962. 'The surface properties of cancer cells: A Review': *Cancer Res.* 22, 525–48.
3. ABERCROMBIE, M., and HEAYSMAN, J. E. M., 1953. 'Social behaviour of cells in tissue culture. I. Speed of movement of chick heart fibroblasts in relation to their mutual contacts': *Exper. Cell Res.* 5, 111–31.
4. ABERCROMBIE, M., and HEAYSMAN, J. E. M., 1954. 'Social behaviour of cells in tissue culture. II. Monolayering of fibroblasts': *Exper. Cell Res.* 6, 293–306.
5. AFZELIUS, B. A., 1955. 'The ultrastructure of the nuclear membrane of the sea urchin oocyte as studied with the electron microscope': *Exper. Cell Res.* 8, 147–58.
6. ALEXANDER, M., BRESLER, A., FURTH, J., and HURWITZ, J., 1960. 'The incorporation of ribonucleotides into RNA': *Fed. Proc.* 19, 318.
7. ALLEN, J. M. (Ed.), 1962. *The molecular control of cellular activity:* McGraw-Hill, New York.
8. ALLFREY, V. G., LITTAU, V. C., and MIRSKY, A. E., 1963. 'On the role of histones in regulating ribonucleic acid synthesis in the cell nucleus': *Proc. Nat. Acad. Sci.* 49, 414–21.
9. ALLFREY, V. G., and MIRSKY, A. E., 1961. 'How cells make molecules': *Scientific American* 205, 74–96.
10. AMANO, S., 1957. 'The structure of the centrioles and spindle body as observed under the electron and phase contrast microscopes. A new extension-fibre theory concerning mitotic mechanism in animal cells': *Cytologia, Int. J. of Cytology* 22, 193.
11. AMBROSE, E. J., 1961. 'The movements of fibrocytes': *Exper. Cell Res.* Suppl. 8, 54–73.
12. AMBROSE, E. J., 1965. 'Cell movements': *Endeavour* 24, No. 91, 27–32.
13. ANDERSON, E., and BEAMS, H. W., 1956. 'Evidence from electron micrographs for the passage of material through pores of the nuclear membrane': *J. Biophys. Biochem. Cytol.* 2, Suppl. 439–43.
14. ANFINSEN, C. B., and REDFIELD, R. R., 1956. 'Protein structure in relation to function and biosynthesis': *Adv. in Protein Chem.* 11, 1–100.

15. ARNOLD, W., and SHERWOOD, H. K., 1957. 'Are chloroplasts semi-conductors?': *Proc. Nat. Acad. Sci.* 43, 105–14.
16. ARNON, D. I., 1959. 'Conversion of light into chemical energy in photo-synthesis': *Nature* 184, 10–21.
17. ARNON, D. I., 1962. 'Photosynthetic phosphorylation and a unified concept of photosynthesis': *Comp. Biochem. Physiol.* 4, 253–79.
18. ASTBURY, W. T., and SAHA, N. N., 1953. 'Structure of algal flagella': *Nature* 171, 280–3.
19. AUERBACH, R., and GROBSTEIN, C., 1958. 'Inductive interaction of embryonic tissues after dissociation and reaggregation': *Exper. Cell Res.* 15, 384–97.
20. AUGUST, J. T., ORTIZ, P. J., and HURWITZ, J., 1962. 'Ribonucleic acid-dependent ribonucleotide incorporation. I. Purification and properties of the enzyme': *J. Biol. Chem.* 237, 3786–93.
21. AVERY, O. T., MCLEOD, C. M., and MCCARTY, M., 1944. 'Studies on the chemical nature of the substance inducing transformation of pneumococcal types': *J. Exper. Med.* 79, 137–57.
22. BAJER, A., 1965. Quoted by Peacock, W. J., 1963, ref. 374.
23. BASILIO, C., WAHBA, A. J., LENGYEL, P., SPEYER, J. F., and OCHOA, S., 1962. 'Synthetic polynucleotides and the amino acid code, V': *Proc. Nat. Acad. Sci.* 48, 613–16.
24. BASSHAM, J. A., 1962. 'The path of carbon in Photosynthesis': *Scientific American* 206, 88–100.
25. BASSHAM, J. A., and CALVIN, M. C., 1957. *The Path of Carbon in Photosynthesis:* Prentice-Hall Inc., Englewood Cliffs, New Jersey.
26. BASSHAM, J. A., and CALVIN, M., 1962. 'The way of CO_2 in plant photosynthesis': *Comp. Biochem. Physiol.* 4, 187–204.
27. BAUDHUIN, P., BEAUFAY, H., and DE DUVE, C., 1965. 'Combined biochemical and morphological study of particulate fractions from rat liver. Analysis of preparations enriched in lysosomes or in particles containing urate oxidase, D-amino acid oxidase, and catalase': *J. Cell Biol.* 26, 219–43.
28. BECKER, U., and TIEDEMANN, H., 1959. 'Versuche zur Determination von embryonalem Amphibiengewebe durch Induktionsstoffe in Lösung': *Zeitschrift fur Naturforschung* 14b, 608–9.
29. BECKER, U., and TIEDEMANN, H., 1961. 'Zell- und Organdetermination in der Gewebekultur, ausgeführt am präsumptiven Ektoderm der Amphibiengastrula': Verhandlungen der Deutschen Zoologischen Gesellschaft in Saarbrücken : Geese & Portig, Leipzig, pp. 259–67.
30. BEERMANN, W., 1952. 'Chromosomenkonstranz und spezifische Modifikationene der Chromosomenstruktur in der Entwicklung und Organ-differenzierung von *Chironomus tentans*': *Chromosoma* 5, 139–98.

31. BEERMANN, W., 1959. 'Chromosomal differentiation in insects': in Rudnick, D. (Ed.), *Developmental Cytology*, pp. 83–103: The Ronald Press, New York.
32. BELL, D. J., and GRANT, J. K. (Eds.), 1962. *The Structure and Biosynthesis of Macromolecules:* Cambridge University Press.
33. BENNETT, H. S., 1956. 'The concepts of membrane flow and membrane vesiculation as mechanisms for active transport and ion pumping': *J. Biophys. Biochem. Cytol.* Suppl. 2, 99–103.
34. BENZER, S. 1957. 'The elementary units of heredity': in McElroy, W., and Glass, B. (Eds.), *Chemical Basis of Heredity:* Johns Hopkins Press, Baltimore.
35. BENZER, S., 1962. 'The fine structure of the gene': *Scientific American* 206, 70–84.
36. BERNFIELD, M. R., and NIRENBERG, M. W., 1965. 'RNA code-words and protein synthesis': *Science* 147, 479–84.
37. BESSIS, M., BRETON-GORIUS, J., and THIERY, J. P., 1958. 'Centriole, corps de Golgi et aster des leucocytes. Etude au microscope electronique'; *Rev. Hemat.* 13, 363–86.
38. BIRBECK, M. S. C., and MERCER, E. H., 1956. 'Cell membranes and morphogenesis': *Nature*, 178, 985–6.
39. BIRNSTIEL, M. L., CHIPCHASE, M. I. H., and HYDE, B. B., 1963. 'The nucleolus, a source of ribosomes': *Biochim. Biophys. Acta* 76, 454.
40. BISHOP, J., ALLEN, E., LEAHY, J., MORRIS, A., and SCHWEET, R., 1960. 'Stages in hemoglobin synthesis in a cell-free system': *Fed. Proc.* 19, 346.
41. BISHOP, J., LEAHY, J., and SCHWEET, R., 1960. 'Formation of the peptide chain of hemoglobin': *Proc. Nat. Acad. Sci.* 46, 1030–8.
42. BLAIR, D. G., and POTTER, V. R., 1961. 'Inhibition of orotidylic acid decarboxylase by uridine-5′-phosphate': *J. Biol. Chem.* 236, 2503–6.
43. BOEZI, J. A., and COWIE, D. B., 1961. 'Kinetic studies of β-galactosidase induction': *Biophys. J.* 1, 639–47.
44. BOIVIN, A., VENDRELY, R., and VENDRELY, C., 1948. 'L'acid desoxyribonucleique du noyau cellulaire, depositaire des caracteres hereditaires': *Compt. rend. Acad. Sci.* 226, 1061–3.
45. BOLLUM, F. J., 1960. 'Calf thymus polymerase': *J. Biol. Chem.* 235, 2399–403.
46. BONNER, J., HUANG, R. C., and GILDEN, R. V., 1963. 'Chromosomally directed protein synthesis': *Proc. Nat. Acad. Sci.* 50, 893–9.
47. BONNER, J. T., 1952. 'The pattern of differentiation in amoeboid slime molds': *Naturalist* 86, 79–89.
48. BONNER, J. T., 1959. 'Evidence for the sorting-out of cells in the development of the cellular slime molds': *Proc. Nat. Acad. Sci.* 45, 379–84.

49. BOOIJ, H. L., and BUNGENBERG DE JONG, H. G., 1956. 'Biocolloids and their interactions': *Protoplasmatologia* 1, 2.
50. BOTTS, J., and MORALES, M. F., 1951. 'The elastic mechanism and hydrogen bonding in actomyosin threads': *J. Cell. Comp. Physiol.* 37, 27–56.
51. BOURNE, C. H. (Ed.), 1960. *The Structure and Function of Muscle*, vols. 1–3: Academic Press, New York.
52. BOWEN, W. J., and MARTIN, H. L., 1962. 'The ability of glycerol-treated muscle fibres to do work during ATP-induced contractions and the free energy of ATP': *Arch. Biochem. Biophys.* 98, 364–373.
53. BRACHET, J., 1957. *Biochemical Cytology:* Academic Press, New York.
54. BRACHET, J., 1960. *The Biological Role of Ribonucleic Acids:* Elsevier, Amsterdam.
55. BRADFIELD, J. R. G., 1955. 'Fibre patterns in animal flagella and cilia': *Symp. Soc. Exp. Biol.* 9, 306–34.
56. BRADFUTE, O. E., CHAPMAN-ANDRESEN, C., and JENSEN, W. A., 1964. 'Concerning morphological evidence for pinocytosis in higher plants': *Exper. Cell Res.* 36, 207–10.
57. BRANDES, D., and BERTINI, F., 1964. 'Role of Golgi apparatus in the formation of cytolysomes': *Exper. Cell Res.* 35, 194–217.
58. BRENNER, S., JACOB, F., and MESELSON, M., 1961. 'An unstable intermediate carrying information from genes to ribosomes for protein synthesis': *Nature* 190, 576–81.
59. BRENNER, S., STRETTON, A. O. W., and KAPLAN, S., 1965. 'Genetic code: the "nonsense" triplets for chain termination and their suppression': *Nature* 206, 994–8.
60. BRENNER, S., and STRETTON, A. O., 1965. 'Phase shifting of *Amber* and *Ochre* mutants': *J. Mol. Biol.* 13, 944–6.
61. BRESNICK, E., 1962. 'Feedback inhibition of aspartate transcarbamylase in liver and in hepatoma': *Cancer Res.* 22, 1246–51.
62. BRESNICK, E., and HITCHINGS, G. H., 1961. 'Feedback control in Ehrlich ascites cells': *Cancer Res.* 21, 105–9.
63. BRETSCHER, M. S., GOODMAN, H. M., MENNINGER, J. R., and SMITH, J. D., 1965. 'Polypeptide chain termination using synthetic polynucleotides': *J. Mol. Biol.* 14, 634–9.
64. BRIERLEY, G. P., BACHMANN, E., and GREEN, D. E., 1962. 'Active transport of inorganic phosphate and magnesium ions by beef heart mitochondria': *Proc. Nat. Acad. Sci.* 4, 1928–35.
65. BRIERLEY, G., and GREEN, David E., 1965. 'Compartmentation of the mitochondrion': *Proc. Nat. Acad. Sci.* 53, 73–9.
66. BRIGGS, R., and King, T. J., 1960. 'Nuclear transplantation studies on the early gastrula (*Rana pipiens*). I. Nuclei of presumptive endoderm': *Dev. Biol.* 2, 252–70.

67. BROOKS, S. C., and BROOKS, M. M., 1941. *The permeability of living cells:* Gebr. Borntraeger, Berlin.
68. BROWN, D. D., and GURDON, J. B., 1964. 'Absence of ribosomal RNA synthesis in the anucleolate mutant of Xenopus Laevis': *Proc. Nat. Acad. Sci.* 51, 139–46.
69. BURNET, F. M., 1961. 'Immunological recognition of self': *Science* 133, 307–11.
70. BURROWS, R., and LAMB, J. F., 1962. 'Sodium and potassium fluxes in cells cultured from chick embryo heart muscle': *J. Physiol.* 162, 510–31.
71. CAIRNS, J., 1963. 'The bacterial chromosome and its manner of replication as seen by autoradiography': *J. Mol. Biol.* 6, 208–13.
72. CAIRNS, J., 1966. 'Autoradiography of HeLa cell INA': *J. Mol. Biol.* 15, 372–3.
73. CALLAN, H. G., and TOMLIN, S. G., 1950. 'Experimental studies on amphibian oocyte nuclei. I. Investigation of the structure of the nuclear membrane by means of the electron microscope': *Proc. Roy. Soc.* B137, 367–78.
74. CALVIN, M., 1962. 'The origin of life on earth and elsewhere': *Persps. Biol. Med.* V, 399–422.
75. CALVIN, M., 1962. 'The path of carbon in photosynthesis': *Science* 135, 879–89.
76. CALVIN, M., and ANDROES, G. M., 1962. 'Primary quantum conversion in photosynthesis': *Science* 138, 867–73.
77. CANELLAKIS, E. S., 1957. 'On the mechanism of incorporation of adenylic acid from adenosine triphosphate into ribonucleic acid by soluble mammalian enzyme systems': *Biochim. Biophys. Acta* 25, 217–18.
78. CARO, L. G., and PALADE, G. E., 1964. 'Protein synthesis, storage, and discharge in the pancreatic exocrine cell. An autoradiographic study': *J. Cell Biol.* 20, 473–96.
79. CHAPEVILLE, F., LIPMAN, F., VON EHRENSTEIN, G., WEISBLUM, B., RAY, W. J., Jr., and BENZER, S., 1962. 'On the role of soluble ribonucleic acid in coding for amino acids': *Proc. Nat. Acad. Sci.* 48, 1086–92.
80. CHARGAFF, E., 1950. 'Chemical specificity of nucleic acids and mechanism of their enzymatic degradation': *Experientia* 6, 201–9.
81. CHARGAFF, E., and DAVIDSON, J. N., 1960. *The Nucleic Acids,* vols. 1–3 (1955): Academic Press, New York.
82. CLARK, B. F. C., and MARCKER, K. A., 1966. 'The role of N-formylmethionyl-sRNA in protein biosynthesis': *J. Mol. Biol.* 17, 394–406.
83. CLAUDE, A., 1954. 'Fine structure of cytoplasm': *8th Symp. Cell Biol.* Leyden, 304.
84. COHEN, G. N., and MONRO, J., 1957. 'Bacterial permeases': *Bact. Rev.* 21, 169–94.

85. COLE, A., 1962. 'A molecular model for biological contractility: Implications in chromosome structure and function': *Nature* 196, 211–14.

86. COLE, K. W., and CURTIS, H. J., 1939. 'Electric impedance of the squid giant axon during activity': *J. Gen. Physiol.* 22, 649–70.

87. COLE, R. J., and DANIELLI, J. F., 1963. 'Nuclear-cytoplasmic interactions in the responses of Amoeba proteus and Amoeba discoides to streptomycin': *Exper. Cell Res.* 29, 199–206.

88. COLLANDER, R., and BÄRLUND, H., 1933. 'Permeabilitätstudien an Chara ceratosphylla. II. Die Permeabilität für Nichtelektrolyte': *Acta. bot. fenn.* 11, 1–114.

89. COLVILL, A. J. E., KANNER, L. C., TOCCHINI-VALENTINI, G. P., SARNAT, M. T., and GEIDUSCHEK, E. P., 1965. 'Asymmetric RNA synthesis *in vitro*: heterologous DNA-enzyme systems; *E. coli* RNA polymerase': *Proc. Nat. Acad. Sci.* 53, 1140–7.

90. CONKLIN, E. G., 1931. 'Centrifuged eggs of Acidians': *J. Exper. Zool.* 60, 2–120.

91. COOK, G. M. W., HEARD, D. H., and SEAMAN, G. V. F., 1961. 'Sialic acids and the electrokinetic charge of the human erythrocyte': *Nature* 191, 44–7.

92. COREY, R. B., and PAULING, L., 1953. 'Fundamental dimensions of polypeptide chains': *Proc. Roy. Soc.*, B141, 10–20.

93. COX, R. P., and MACLEOD, C. M., 1962. 'Alkaline phosphatase content and the effects of prednisolone on mammalian cells in culture': *J. Gen. Physiol.* 45, 439–85.

94. COX, R. P., and MACLEOD, C. M., 1963. 'Repression of alkaline phosphatase in human cell cultures by cystine and cysteine': *Proc. Nat. Acad. Sci.* 49, 504–10.

95. COX, R. P., and PONTECORVO, G., 1961. 'Induction of alkaline phosphatase by substrates in established cultures of cells from individual human donors': *Proc. Nat. Acad. Sci.* 47, 839–45.

96. CRICK, F. H. C., 1958. 'On protein synthesis': *Symp. Soc. Exp. Biol.* 12, 138–63.

97. CRICK, F. H. C., 1963. 'On the genetic code': *Science* 139, 461–4.

98. CRICK, F. H. C., BARNETT, L., BRENNER, S., and WATTS-TOBIN, R. J., 1961. 'General nature of the genetic code for proteins': *Nature* 192, 1227–32.

99. CRIDDLE, R. S., BOCK, R. M., GREEN, D. E., and TISDALE, H., 1962. 'Physical characteristics of proteins of the electron transfer system and interpretation of the structure of the mitochondrion': *Biochemistry* 1, 827–42.

100. CROUSE, H. V., 1954. X-ray breakage of chromosomes at first meiotic metaphase': *Science* 119, 485–7.

101. CUNNINGHAM, W. P., MORRÉ, D. J., and MOLLENHAUER, H. H., 1966. 'Structure of isolated plant Golgi apparatus revealed by negative staining': *J. Cell Biol.* 28, 169–79.
102. CURTIS, A. S. G., 1960. 'Cortical grafting in *Xenopus laevis*': *J. Emb. Exp. Morph.* 8, 163–73.
103. CURTIS, A. S. G., 1960. 'Cell contacts: some physical considerations': *Amer. Nat.* 94, 37–56.
104. CURTIS, A. S. G., 1962. 'Cell contact and adhesion': *Biol. Rev.* 37, 82–129.
105. DANIELLI, J., 1938. 'Protein films at the oil–water interface': *Cold Spr. Har. Symp.* 6, 190–5.
106. DANIELLI, J. F., 1954. 'The present position in the field of facilitated diffusion and selective active transport': *Colston Papers* 7, 1–14.
107. DANIELLI, J. F., and HARVEY, E. N., 1934. 'The tension at the surface of mackerel egg oil, with remarks on the nature of the cell surface': *J. Cell. Comp. Physiol.* 5, 483–94.
108. DANIELLI, J. F., PANKHURST, K. G. A., and RIDDIFORD, A. C. (Eds.), 1958. *Surface Phenomena in Chemistry and Biology:* Pergamon Press, London.
109. DAVIDSON, E. H., ALLFREY, V. G., and MIRSKY, A. E., 1963. 'Gene expression in differentiated cells': *Proc. Nat. Acad. Sci.* 49, 53–60.
110. DAVIDSON, J. N., 1960. *The Biochemistry of the Nucleic Acids*, 4th edn.: Methuen, London.
111. DAVIDSON, J. N., 1962. 'The control of DNA biosynthesis': from *The Molecular Basis of Neoplasia:* University Texas Press, Austin.
112. DAVSON, H., and DANIELLI, J. F., 1943. In *The Permeability of Natural Membranes:* Cambridge University Press, London.
113. DAWSON, R. M. C., HEMIGTON, N., and LINDSAY, D. B., 1960. 'The phospholipids of the erythrocyte "Ghosts" of various species': *Biochem. J.* 77, 226–30.
114. DE BERNARD, B., 1962. 'The respiratory chain in animal mitochondria': *Ital. J. Biochem.* 11, 1–41.
115. DE DUVE, C., 1959. 'Lysosomes, a new group of cytoplasmic particles': in Hayashi, T. (Ed.), *Subcellular Particles:* The Ronald Press, New York.
116. DE MARS, R., 1958. 'The inhibition by glutamine of glutamyltransferase formation in human cells in culture': *Biochim. Biophys. Acta* 27, 435–6.
117. DE ROBERTIS, E. D. P., NOWINSKY, W. W., and SAEZ, F. A., 1960. *General cytology*, 3rd edn.: Saunders, Philadelphia.
118. DERVICHIAN, D. G., 1958. 'The existence and significance of molecular association in monolayers': in Danielli, J. F., Pankhurst, K. G. A., and Riddiford, A. C. (Eds.), *Surface Phenomena in Chemistry and Biology:* Pergamon Press, London.
119. DE-THÉ, G., 1964. 'Cytoplasmic microtubules in different animal cells': *J. Cell Biol.* 23, 265–76.

120. DOUNCE, A. L., and SARKAR, N. K., 1960. 'Nucleoprotein organization in cell nuclei and its relationship to chromosomal structure': in Mitchell (Ed.), *The Cell Nucleus:* Academic Press, New York.
121. EAKIN, R. E., 1963. 'An approach to the evolution of metabolism': *Proc. Nat. Acad. Sci.* 49, 360–6.
122. EASTY, G. C., and MUTOLO, V., 1960. 'The nature of the intercellular material of adult mammalian tissues': *Exper. Cell Res.* 21, 374–85.
123. ENGEL, L. L., and SCOTT, J. F., 1960. 'Effects of steroid hormones upon diphosphopyridine nucleotide mediated enzymatic reactions': *Rec. Prog. Hormone Res.* 16, 79–91.
124. ENGSTRÖM, A., and FINEAN, J. B., 1958. *Biological Ultrastructure:* Academic Press, New York.
125. ERRERA, M., HELL, A., and PERRY, R. P., 1961. 'The role of the nucleolus in ribonucleic acid and protein synthesis. II. Amino acid incorporation into normal and nucleolar inactivated HeLa cells': *Biochim. Biophys. Acta* 49, 58–63.
126. FAURE-FREMIET, E., 1957. 'Finer morphology of microorganisms': *Ann. Rev. Microbiol.* 11, 1–6.
127. FAURE-FREMIET, E., 1958. 'Structure et ultra-structure des protistes': *Rev. Path. Gen.* 695, 265–81.
128. FAWCETT, D. W., 1956. 'The fine structure of chromosomes in the meiotic prophase of vertebrate spermatocytes': *J. Biophys. Biochem. Cytol.* 2, 403–6.
129. FAWCETT, D. W., 1958. 'Structural specializations of the cell surface': in Palay, S. L. (Ed.), *Frontiers in Cytology*, pp. 19–41: Yale University Press, New Haven.
130. FAWCETT, D. W., 1960. 'The fine structure of small blood vessels': in *The Microcirculation:* University of Illinois Press, Urbana.
131. FAWCETT, D. W., 1961. 'Intercellular bridges': *Exper. Cell Res.* Suppl. 8, 174–87.
132. FAWCETT, D. W., 1961. 'The membranes of the cytoplasm': *Lab. Invest.* 10, 1162–88.
133. FAWCETT, D. W., and ITO, S., 1958. 'Observations on the cytoplasmic membranes of testicular cells examined by phase contract and electron microscopy': *J. Biophys. Biochem. Cytol.* 4, 135–41.
134. FAWCETT, D. W., and PORTER, K. R., 1952. 'A study of the fine structure of ciliated epithelial cells with the electron microscope': *Anat. Rec.* 113, 539.
135. FAWCETT, D. W., and PORTER, K. R., 1954. 'A study of the fine structure of ciliated epithelia': *J. Morph.* 94, 221–81.
136. FELDHERR, C. M., 1962. 'The nuclear annuli as pathways for nucleocytoplasmic exchanges': *J. Cell. Biol.* 14, 65–72.
137. FELIX, K., FISCHER, H., and KREKELS, A., 1956. 'Protamines and nucleoprotamines': *Progress in Biophys.* 6, 2–23.

138. FELL, H. B., 1961. 'Changes in synthesis induced in organ cultures': from *Molecular and Cellular Synthesis, 19th Growth Symposium*, The Ronald Press.

139. FELL, H. B., 1962. 'Some effects of environment on epidermal differentiation': *Brit. J. Dermatol.* 74, 1–7.

140. FERNANDEZ-MORAN, H., 1954. 'The submicroscopic structure of nerve fibers': *Progress in Biophys. & Biophys. Chem.* 4, 112–47.

141. FERNANDEZ-MORAN, H., 1962. 'Low-temperature electron microscopy and X-ray diffraction studies of lipoprotein components in lamellar systems': *Circulation* 26, 1039–65.

142. FERNANDEZ-MORAN, H., and FINEAN, J. B., 1957. 'Electron microscope and low-angle X-ray diffraction studies of the nerve myelin sheath': *J. Biophys. Biochem. Cytol.* 3, 725–48.

143. FICQ, A., and PAVAN, C., 1957. 'Autoradiography of polytene chromosomes of *Rhynchosciara angelae* at different stages of larval development': *Nature* 180, 983–4.

144. FINEAN, J. B., 1953. 'Further observations on the structure of myelin': *Exper. Cell Res.* 5, 202–15.

145. FINEAN, J. B., 1961. *Chemical Ultrastructure in Living Tissues.* Thomas, Springfield, Illinois.

146. FISCHBERG, M., and BLACKLER, A. W., 1961. 'How cells specialize': *Scientific American* 205, 124–40.

147. FISCHBERG, M., GURDON, J. G., and ELSDALE, T. R., 1958. 'Nuclear transfer in Amphibia and the problem of the potentialities of the nuclei of differentiating tissues': *Exper. Cell. Res.* Suppl. 6, 161–78.

148. FLEISSNER, E., and BOREK, E., 1962. 'A new enzyme of RNA synthesis: RNA methylase': *Proc. Nat., Acad. Sci.* 48, 1199–203.

149. FLORKIN, M. (Ed.), 1960. *Aspects of the origin of life:* Pergamon Press, London.

150. FLORY, P. J., 1953. *Principles of Polymer Chemistry:* Cornell University Press, Ithaca, New York.

151. FOWLER, L. R., and RICHARDSON, S. H., 1963. 'Studies on the electron transfer system. I. On the mechanism of reconstitution of the mitochondrial electron transfer system': *J. Biol. Chem.* 238, 456–63.

152. FREESE, E., 1961. 'The Molecular Mechanisms of Mutations': *Proc. Vth Int. Congr. Biochem.* Symp. 1, Pergamon Press, London.

153. FREY-WYSSLING, A., 1953. *Submicroscopic Morphology of Protoplasm*, 2nd ed.: Elsevier, Amsterdam.

154. FRIEDRICH-FRESKA, H., and HANDEWITZ, F., 1953. 'Letale Spatfolgen nach Einbau von P^{32} in Amoeba proteus und ihre Deutung durch genetische Untereinheiten': *Z. Naturforsch.* 8, 343–55.

155. GALL, J. G., 1956. 'On the submicroscopic structure of chromosomes': *Brookhaven Symposia in Biol.* 8, 17–32.

156. GALL, J. G., 1959. 'The nuclear envelope after $KMnO_4$-fixation': *J. Biophys. Biochem. Cytol.* 6, 115–17.
157. GARDNER, R. S., WAHBA, A. J., BASILIO, C., MILLER, R. S., LENGYEL, P., and SPEYER, J. F., 1962. 'Synthetic polynucleotides and the amino acid code, VII': *Proc. Nat. Acad. Sci.* 48, 2087–94.
158. GAREN, A., and ECHOLS, H., 1962. 'Genetic control of induction of alkaline phosphatase synthesis in *E. coli*': *Proc. Nat. Acad. Sci.* 48, 1398–1402.
159. GEIDUSCHEK, E. P., NAKAMOTO, T., and WEISS, S. B., 1961. 'The enzymatic synthesis of RNA: complementary interaction with DNA': *Proc. Nat. Acad. Sci.* 47, 1405–15.
160. GENEVES, L., LANCE, A., and BUVAT, R., 1958. 'Sur la presence dans le cytoplasme vegetal, et sur la nature ergastoplasmique constituants figures analogues aux "lysosomes" ou aux "dense bodies" des cellules animales': *Compt. Rend. Acad. Sci.* 247, 2028.
161. GEORGIEV, G. P., and MANTIEVA, V. L., 1960. 'Isolation of cell nuclei by means of a phenol method and their characteristics': *Biokhimiya* 25, 143–50 (original in Russian).
162. GEORGIEV, G. P., and MANTIEVA, V. L., 1962. 'The isolation of DNA-like RNA and ribosomal RNA from the nucleolochromosomal apparatus of mammalian cells': *Biochim. Biophys. Acta* 61, 153–4.
163. GEREN, B. B., and SCHMIDT, F. O., 1953. 'The structure of the nerve sheath in relation to lipid and lipid-protein layers': *J. App. Physics.* 24, 1421.
164. GERHART, J. C., and PARDEE, A. B., 1962. 'The enzymology of control by feedback inhibition': *J. Biol. Chem.* 237, 891–6.
165. GERSHFIELD, N. L., and HEFTMANN, E., 1963. 'Steroid hormones and monolayers': *Experientia* 19, 1–2.
166. GIBBONS, I. R., 1965. 'Reactivation of glycerinated cilia from *Tetrahymena pyriformis*': *J. Cell Biol.* 25, 400–2.
167. GLYNN, I. M., 1956. 'Sodium and potassium movements in human red cells': *J. Physiol.* 134, 278–310.
168. GOLDACRE, R. J., and BEAN, A. D., 1960. 'A model for morphogenesis': *Nature* 186, 294–5.
169. GOLDACRE, R. J., and LORCH, I. J., 1950. 'Folding and unfolding of protein molecules in relation to cytoplasmic streaming, amoeboid movement and osmotic work': *Nature* 166, 497–500.
170. GOLDSTEIN, L., and MICOU, J., 1959. 'Nuclear cytoplasmic relationships in human cells in tissue culture. III. Auto-radiographic study of interrelation of nuclear and cytoplasmic ribonucleic acid': *J. Biophys. Biochem. Cytol.* 6, 1–6.
171. GOLDSTEIN, L., and PLAUT, W., 1955. 'Direct evidence for nuclear synthesis of cytoplasmic ribose nucleic acid': *Proc. Nat. Acad. Sci.* 41, 874–80.

172. GOODMAN, H. M., and RICH, A., 1962. 'Formation of a DNA-soluble RNA hybrid and its relation to the origin, evolution and degeneracy of soluble RNA': *Proc. Nat. Acad. Sci.* 48, 2101–8.

173. GORDON, G. B., MILLER, L. R., and BENSCH, K. G., 1965. 'Studies on the intracellular digestive process in mammalian tissue culture cells': *J. Cell Biol.* 25, 41–55.

174. GORINI, L., and MAAS, W., 1958. 'Feedback control of the formation of biosynthetic enzymes': in McElroy, W. T., and Glass, B. (Eds.), *Chemical Basis of Development:* Johns Hopkins Press, Baltimore, p. 469.

175. GRASSE, P. P., and CARASSO, N., 1957. 'Ultrastructure of the Golgi apparatus in Protozoa and Metazoa (somatic and germinal cells)': *Nature* 179, 31–3.

176. GRASSE, P. P., CARASSO, N., and FAVARD, P., 1955. 'Les dictyosomes (appareil de Golgi) et leur ultra-structure': *Compt. Rend. Acad. Sci.* 241, 1243–5.

177. GREEN, B. B., 1959. 'The formation from the Schwann cell surface of myelin in the peripheral nerves of chick embryos': *Exper. Cell Res.* 7, 558–62.

178. GREEN, D. E., 1962. 'Structure and function of subcellular particles': *Comp. Biochem. Physiol.* 4, 81–122.

179. GREEN, D. E., TISDALE, H. D., CRIDDLE, R. S., and BOCK, R. M., 1961. 'The structural protein and mitochondrial organisation': *Biochem. Biophys. Res. Comm.* 5, 81–4.

180. GREENGARD, O., SMITH, M. A., and ACS, G., 1963. 'Relation of cortisone and synthesis of ribonucleic acid to induced and developmental enzyme formation': *J. Biol. Chem.* 238, 1548–51.

181. GRIFFITH, F., 1928. 'The significance of pneumococcal types': *J. Hyg.* 27, 113–59.

182. GROBSTEIN, C., 1954. 'Tissue interaction in the morphogenesis of mouse embryonic rudiments in vitro': in Rudnick, D. (Ed.), *Aspects of Synthesis and Order in Growth:* Princeton University Press, Princeton, New Jersey.

183. GROBSTEIN, C., 1957. 'Kidney tubule induction in mouse metanephrogenic mesenchyme without cytoplasmic contact': *J. exp. Zool.* 135, 57–66.

184. GROBSTEIN, C., 1962. 'Interactive processes in cytodifferentiation': *J. Cell. Comp. Physiol.* Suppl. 1, 60, 35–48.

185. GROS, F., HIATT, H., GILBERT, W., KURLAND, C. G., RISEBROUGH, R. W., and WATSON, J. D., 1961. 'Unstable RNA revealed by pulse labelling of *E. coli*': *Nature* 190, 581–5.

186. GROVER, J. W., 1961. 'The enzymatic dissociation and reproducible reaggregation *in vitro* of 11-day embryonic chick lung': *Dev. Biol.* 3, 555–68.

187. GRUNBERG-MANAGO, M., and OCHOA, S., 1955. 'Enzymatic synthesis and breakdown of polynucleotides: polynucleotide phosphorylase': *J. Amer. Chem. Soc.* 77, 3165–6.

188. GUEST, J. R., and YANOFSKY, C., 1966. 'Relative orientation of gene, messenger and polypeptide chain': *Nature* 210, 799–802.

189. GURDON, J. B., 1962. 'The developmental capacity of nuclei taken from intestinal epithelium cells of feeding tadpoles': *J. Embryol. Exper. Morph.* 10, 622–40.

190. GURDON, J. B., 1962. 'Adult frogs derived from the nuclei of single somatic cells': *Dev. Biol.* 4, 256–73.

191. GUSTAVSON, K. H., 1956. *The Chemistry of Collagen:* Academic Press, New York.

192. HALVORSON, H. O., and GORMON, J., 1959. 'The abnormal pattern of protein synthesis in Pseudomonas azotogenesis in the presence of hexetidine': *Exper. Cell Res.* 17, 522–4.

193. HÄMMERLING, J. Von, 1946. 'Dreikernige Transplantate zwischen Acetabularia crenulata und mediterranea I': *Z. Naturforschung* 1, 337–42.

194. HÄMMERLING, J. Von, 1953. 'Nucleo-cytoplasmic relationships in the development of Acetabularia': *Int. Rev. Cytol.* 2, 475–98.

195. HANSON, J., and HUXLEY, H. E., 1955. 'The structural basis of contraction in striated muscle': *Symp. Soc. Exper. Biol.* No. 16, 228–64.

196. HARRIS, E. J., 1960. *Transport and Accumulation in Biological Systems*, 2nd ed.: Butterworth, London.

197. HARRIS, E. J., and MAIZELS, M., 1951. 'The permeability of human erythrocytes to sodium': *J. Physiol.* 113, 506–24.

198. HARRIS, R. J. C., 1961. *Cell Movement and Cell Contact:* Experimental Cell Research Supplement 8, Academic Press, New York.

199. HARRIS, R. J. C. (Ed.), 1961. *Protein Biosynthesis:* Academic Press, London.

200. HARVEY, E. H. A., 1931. 'Determination of the tension at the surface of the sea urchin': *Biol. Bull.* 60, 67–71.

201. HATEFI, Y., HAAVIK, A. G., FOWLER, L. R., and GRIFFITHS, D. E., 1962. 'Studies on the electron transfer system. XLII. Reconstitution of the electron transfer system': *J. Biol. Chem.* 237, 2661–9.

202. HAWKINS, S. E., and COIE, R. J., 1965. 'Studies on the basis of cytoplasmic inheritance in amoebae': *Exper. Cell Res.* 37, 26–38.

203. HAYASHI, M., and SPIEGELMAN, S., 1961. 'The selective synthesis of informational RNA in bacteria': *Proc. Nat. Acad. Sci.* 47, 1564–80.

204. HAYASHI, T., 1961. 'How cells move': *Scientific American* 205, 184–204.

205. HENDERSON, J. F., 1962. 'Feedback inhibition of purine biosynthesis in ascites tumor cells': *J. Biol. Chem.* 237, 2631–5.

206. HENSHAW, E. C., REVEL, M., and HIATT, H. H., 1965. 'A cytoplasmic particle bearing messenger RNA in rat liver': *J. Mol. Biol.* 14, 241–56.

207. Herbert, E., 1958. 'The incorporation of adenine nucleotides into ribonucleic acid of cell-free systems from liver': *J. Biol. Chem.* 231, 975–86.
208. Hershey, A. D., 1952. 'Reproduction of bacteriophage': *Intern. Rev. Cyt.* 1, 119–34.
209. Hershey, A. D., 1954. 'Conservation of nucleic acids during bacterial growth': *J. Gen. Physiol.* 38, 145–8.
210. Hershey, A. D., 1957. 'Bacteriophages as genetic and biochemical systems': *Advances in Virus Research* 4, 25–62.
211. Hershey, A. D., and Chase, M., 1952. 'Independent functions of viral protein and nucleic acid in growth of bacteriophage': *J. Gen. Physiol.* 36, 39–56.
212. Hess, K., and Kiessig, H., 1948. 'Röntgeninterfernzen von Mischkrystallen in Seifen und von Mischmizellen in Seifenlösungen; zur Anordnung der Enden verschieden langer Paraffinketten in gittergeordnetem Zustand': *Chem. Berichte* 81, 327–40.
213. Hill, R., and Whittingham, C. D., 1955. *Photosynthesis:* Methuen, London.
214. Hoagland, M. B., Zamecnik, P. C., and Stephenson, M. L., 1957. 'Intermediate reactions in protein biosynthesis': *Biochim. Biophys. Acta* 24, 215–16.
215. Hodge, A. J., Huxley, H. E., and Spiro, D., 1954. 'Electron microscope studies on ultra thin sections of muscle': *J.Exper. Med.*99,201–6.
216. Hodge, A. J., McLean, J. D., and Mercer, F. V., 1955. 'Ultrastructure of the lamellae and grana in the chloroplasts of Zea Mays': *J. Biophys. Biochem. Cytol* 1, 605–15.
217. Hoelzl Wallach, D. F., and Eylar, E. H., 1961. 'Sialic acid in the cellular membranes of Ehrlich ascites-carcinoma cells': *Biochim. Biophys. Acta* 52, 594–6.
218. Hoffmann-Berling, H., 1954. 'Adenosintriphosphat als Betriebsstoff von Zellbewegungen': *Biochim. Biophys. Acta* 14, 187–94.
219. Hoffmann-Berling, H., 1955. 'Geisselmodelle und Adenosintriphosphat (ATP)': *Biochim. Biophys. Acta* 16, 146–54.
220. Hokin, L. E., and Hokin, M. R., 1959. 'Evidence for phosphatidic acid as the sodium carrier': *Nature* 184, 1068–9.
221. Holley, R. W., 1966. 'The nucleotide sequence of a nucleic acid': *Scientific American* 214, No. 2, 30–9.
222. Holley, R. W., Apgar, J., Everett, G. A., Madian, J. T., Marquisee, M., Merrill, S. H., Penswick, J. R., and Zamir, A., 1965. 'Structure of a ribonucleic acid': *Science* 147, 1462–5.
223. Holter, H., 1961. 'How things get into cells': *Scientific American* 205, 167–80.
224. Holter, H., and Marshall, J. M., Jr., 1954. 'Studies on pinocytosis in the amoeba *Chaos Chaos*': *Compt. rend trav. lab. Carlsberg Ser Chem.* 29, 7–26.

225. Holtfreter, J., 1943. 'The properties and functions of the surface coat in amphibian embryos': *J. Exper. Zool.* 93, 251–323.
226. Holtfreter, J. A., 1948. 'Significance of the cell membrane in embryonic processes': *Ann. N.Y. Acad. Sci.* 49, 709–60.
227. Holtzer, H., Abbot, J., Lash, J., and Holtzer, S., 1960. 'The loss of phenotypic traits by differentiated cells in vitro. I. Dedifferentiation of cartilage cells': *Proc. Nat. Acad. Sci.* 46, 1533–42.
228. Hommes, F. A., Van Leeuwen, G., and Zilliken, F., 1962. 'Induction of cell differentiation. II. The isolation of a chondrogenic factor from embryonic chick spinal cords and notochords': *Biochim. Biophys. Acta* 56, 320–25.
229. Horiuchi, T., Horiuchi, S., and Mizuno, D., 1955. 'A possible negative feedback phenomenon controlling formation of alkaline phosphomonoesterase in Escherichia coli': *Nature* 183, 1529–30.
230. Hotchkiss, R. D., 1959. 'Function and limitation of deoxyribonucleic acids as genetic determinant': *Proc. Can. Cancer Res. Conf.* 3, 3–12.
231. Huang, C., and Thompson, T. E., 1965. 'Properties of lipid bilayer membranes separating two aqueous phases: determination of membrane thickness': *J. Mol. Biol.* 13, 183–93.
232. Huang, R. C., and Bonner, J., 1962. 'Histone, a suppressor of chromosomal RNA synthesis': *Proc. Nat. Acad. Sci.* 48, 1216–22.
233. Hughes, A. F. W., 1952. *The Mitotic Cycle:* Butterworth, London.
234. Hughes, D. E., 1962. 'The bacterial cytoplasmic membrane': *J. Gen. Microbiol.* 29, 39–46.
235. Hurwitz, J., 1959. 'The enzymatic incorporation of ribonucleotides into polydeoxynucleotide material': *J. Biol. Chem.* 234, 2351–8.
236. Hurwitz, J., and Furth, J. J., 1962. 'Messenger RNA': *Scientific American* 206, 41–9.
237. Huxley, A. F., 1957. 'Muscle structure and theories of contraction': *Prog. Biophys.* 7, 255–318.
238. Huxley, A. F., and Huxley, H. E., 1964. 'A discussion of the physical and chemical basis of muscular contraction': *Proc. Roy. Soc.* B160, 433–542.
239. Huxley, A. F., and Taylor, R. E., 1958. 'Local activation of striated muscle fibers': *J. Physiol.* 144, 426–41.
240. Huxley, H. E., 1965. 'The mechanism of muscular contraction': *Scientific American* 213, No. 6, 18–27.
241. Imamoto, F., Morikawa, N., and Sato, K., 1965. 'On the transcription of the tryptophan operon in *Escherichia coli*. III. Multicistronic messenger RNA and polarity for transcription': *J. Mol. Biol.* 13, 169–82.
242. Ingram, V. M., 1961. 'Gene evolution and the haemoglobins': *Nature* 189, 704–8.
243. Ingram, V. M., 1961. *Hemoglobin and its abnormalities:* Thomas, Springfield, Illinois.

244. INGRAM, V. M., 1962. 'The evolution of a protein': *Fed. Proc.* 21, 1053–7.
245. IVES, D. H., MORSE, P. A., Jr., and POTTER, V. R., 1963. 'Feedback inhibition of thymidine kinase by thymidine triphosphate': *J. Biol. Chem.* 238, 1467–74.
246. JACOB, F., and MONOD, J., 1961. 'Genetic regulatory mechanism in the synthesis of protein': *J. Mol. Biol.* 3, 318–56.
247. JAKUS, M. A., and HALL, C. E., 1947. 'Studies of actin and myosin': *J. Biol. Chem.* 167, 705–14.
248. JAMIESON, J. D., and PALADE, G. E., 1966. 'Role of the Golgi complex in the intracellular transport of secretory proteins': *Proc. Nat. Acad. Sci.* 55, 424–31.
249. JOKLIK, W. K., and BECKER, Y., 1965. 'Studies on the genesis of polyribosomes. I. Origin and significance of the subribosomal particles.' *J. Mol. Biol.* 13, 496–510.
250. JOKLIK, W. K., and BECKER, Y., 1965. 'Studies on the genesis of polyribosomes. II. The association of nascent messenger RNA with the 40s subribosomal particle': *J. Mol. Biol.* 13, 511–20.
251. JOLY, M., 1950. 'General theory of the structure, transformations and mechanical properties of monolayers': *J. Coll. Sci.* 5, 49–70.
252. JONES, D. S., NISHIMURA, S., and KHORANA, H. G., 1966. 'Studies on polynucleotides. LVI. Further synthesis, *in vitro*, of copolypeptides containing two amino acids in alternating sequence dependent upon DNA-like polymers containing two nucleotides in alternating sequence': *J. Mol. Biol.* 16, 454–72.
253. JONES, O. W., Jr., and NIRENBERG, M. W., 1962. 'Qualitative survey of RNA codewords': *Proc. Nat. Acad. Sci.* 48, 2115–23.
254. JUKES, T. H., 1962. 'Relations between mutations and base sequences in the amino acid code': *Proc. Nat. Acad. Sci.* 48, 1809–15.
255. JUKES, T. H., 1963. 'The genetic code': *American Scientist* 51, 227–45.
256. KAJI, H., and KAJI, A., 1964. 'Specific binding of sRNA with the template-ribosome complex': *Proc. Nat. Acad. Sci.* 52, 1541–7.
257. KANO-SUEOKA, T., and SPIEGELMAN, S., 1962. 'Evidence for a nonrandom reading of the genome': *Proc. Nat. Acad. Sci.* 48, 1942–9.
258. KANNO, Y., ASHMAN, R. F., and LOEWENSTEIN, W. R., 1965. 'Nucleus and cell membrane conductance in marine oocytes': *Exper. Cell Res.* 39, 184–9.
259. KARAKASHIAN, M. W., and WOODLAND HASTINGS, J., 1962. 'The inhibition of a biological clock by actinomycin D': *Proc. Nat. Acad. Sci.* 48, 2130–7.
260. KARREMAN, G., and STEELE, R. H., 1957. 'On the possibility of long distance energy transfer by resonance in biology': *Biochim. Biophys. Acta* 25, 280–91.

261. KAUFMANN, B. P., GAY, H., and McDONALD, M., 1960. 'Organizational patterns within chromosomes': *Intern. Rev. Cytol.* 9, 77–127.
262. KAUFMANN, B. P., and McDONALD, M. R., 1957. 'The organization of the chromosome': *Cold. Spr. Harbor Symp. Quant. Biol.* 21, 233–6.
263. KELLENBERGER, E., 1960. 'The physical state of the bacterial nucleus': in Hayes and Clowes (Eds.), *Microbial Genetics*, pp. 39–46: Cambridge University Press.
264. KELLER, E. B., and ZAMECNICK, P. C., 1956. 'The effect of guanosine diphosphate and triphosphate on the incorporation of labeled amino acids into proteins': *J. Biol. Chem.* 221, 45–9.
265. KENDREW, J. C., 1961. 'The structure of globular proteins': in Goodwin, T. W., and Lindberg, O. (Eds.), *Biological structure and function:* Academic Press, London.
266. KENDREW, J. C., BOBO, G., DINTZIS, H. M., PARRISH, R. C., WYCKOFF, H., and PHILLIPS, D. C., 1958. 'A three-dimensional model of the myoglobin molecule obtained by X-ray analysis': *Nature* 181, 662–6.
267. KENDREW, J. C., DICKERSON, R. E., STRANDBERG, B. E., HART, R. G., DAVIES, D. R., PHILLIPS, D. C., and SHORE, V. C., 1960. 'Structure of myglobin: a three-dimensional Fourier synthesis at 2 Å resolution': *Nature* 185, 422–7.
268. KING, T. J., and BRIGGS, R., 1955. 'Changes in the nuclei of differentiating gastrula cells, as demonstrated by nuclear transplantation': *Proc. Nat. Acad. Sci.* 41, 321–5.
269. KLEIN, E., 1961. 'Studies on the substrate-induced arginase synthesis in animal cell strains cultured *in vitro*': *Exper. Cell Res.* 22, 226–32.
270. KNOX, W. E., and AUERBACH, V. R., 1955. 'The hormonal control of tryptophan peroxidase in the rat': *J. biol. Chem.* 214, 307–13.
271. KONIJN, T. M., and RAPER, K. B., 1961. 'Cell aggregation in *Dictyostelium discoideum*': *Rev. Biol.* 3, 725–756.
272. KORNBERG, A., 1960. 'Biologic synthesis of deoxyribonucleic acid': *Science* 131, 1503–8.
273. KORNBERG, A., LEHMAN, I. R., and SIMMS, E. S., 1956. 'Polydesoxyribonucleotide synthesis by enzymes from *Escherichia coli.*': *Fed. Proc.* 15, 291–2.
274. KREBS, H. A., and KORNBERG, A. L., 1957. *Energy Transformations in Living Matter:* Springer-Verlag, Berlin.
275. KROEGER, H., 1960. 'The induction of new puffing patterns by transplantation of salivary gland nuclei into egg cytoplasm of *Drosophila*': *Chromosoma* 11, 129–45.
276. KROEGER, H., and LEZZI, M., 1966. 'Regulation of gene action in cell development': *Ann. Rev. Entomol.* 11, 1–22.

277. LASH, J. W., 1963. 'Studies on the ability of embryonic mesonephros explants to form cartilage': *Dev. Biol.* 6, 219–32.

278. LASH, J. W., HOMMES, F. A., and ZILLIKEN, F., 1962. 'Induction of cell differentiation. I. The in vitro induction of vertebral cartilage with a low-molecular weight tissue component': *Biochim. Biophys. Acta* 56, 313–19.

279. LATHAM, H., and DARNELL, J. E., 1965. 'Entrance of mRNA into HeLa cell cytoplasm in puromycin-treated cells': *J. Mol. Biol.* 14, 13–22.

280. LEDER, P., and NIRENBERG, M., 1964. 'RNA codewords and protein synthesis. II. Nucleotide sequence of an RNA codeword'; *Proc. Nat. Acad. Sci.* 52, 420–6.

281. LEDER, P., and NIRENBERG, M. W., 1964. 'RNA codewords and protein synthesis. III. On the nucleotide sequence of a cysteine and a leucine RNA codeword': *Proc. Nat. Acad. Sci.* 52, 1521–9.

282. LEDBETTER, M. C., and PORTER, K. K., 1963. 'A "microtubule" in plant cell fine structure': *J. Cell Biol.* 19, 239–50.

283. LEHMAN, I. R., ZIMMERMAN, S. B., ADLER, J., BESSMAN, M. J., SIMMS, E. S., and KORNBERG, N., 1959. 'Enzymatic synthesis of deoxyribonucleic acid. V. Chemical composition of enzymatically synthesized deoxyribonucleic acid': *Proc. Nat. Acad. Sci.* 44, 1191–6.

284. LEHNINGER, A. L., 1961. 'How cells transform energy': *Scientific American* 205, 62–73.

285. LEHNINGER, A. L., 1964. *The mitochondrion.* Benjamin, New York.

286. LEHNINGER, A. L., WADKINS, C. L., COOPER, C., DEVLIN, T. M., and GAMBLE, J. L., Jr., 1958. 'Oxidative phosphorylation': *Science* 12, 450–6.

287. LENGYEL, P., SPEYER, J. F., BASILIO, C., and OCHOA, S., 1962. 'Synthetic polynucleotides and the amino acid code, III': *Proc. Nat. Acad. Sci.* 48, 282–4.

288. LENGYEL, P., SPEYER, J. F., and OCHOA, S., 1961. 'Synthetic polynucleotides and the amino acid code': *Proc. Nat. Acad. Sci.* 47, 1936–42.

289. LERMAN, L. S., 1963. 'The structure of the DNA-acridine complex': *Proc. Nat. Acad. Sci.* 49, 94–102.

290. LESLIE, I., 1961. 'Biochemistry of heredity: a general hypothesis': *Nature* 189, 260–8.

291. LEVI-MONTALCINI, R., and ANGELETTI, P. V., 1962. 'Growth and differentiation': *Ann. Rev. Physiol.* 24, 11–56.

292. LEVINTHAL, C., 1956. 'The mechanism of DNA replication and genetic recombination in phage': *Proc. Nat. Acad. Sci.* 42, 394–404.

293. LEVINTHAL, C., KEYMAN, A., and HIGA, A., 1962. 'Messenger RNA turnover and protein synthesis in B. subtilis inhibited by Actinomycin D': *Proc. Nat. Acad. Sci.* 48, 1631–8.

294. LEVY, H. M., and KOSHLAND, D. E., Jr., 1959. 'Mechanism of hydrolysis of adenosine triphosphate by muscle proteins and its relation to muscular contraction': *J. Biol. Chem.* 234, 1102–7.

295. LOWENSTEIN, W. R., SOCOLAR, S. J., HIGASHINO, S., KANNO, Y., and DAVIDSON, N., 1965. 'Intercellular communication: renal, urinary bladder, sensory, and salivary gland cells': *Science* 149, 295–8.

296. LWOFF, A., 1943. *L'evolution physiologique. Etude des pertes de fonction chez les microorganismes:* Hermann, Paris.

297. MCCLINTOCK, B., 1934. 'The relation of a particular chromosomal element to the development of the nucleoli in *Zea mays*': *Z. Mikr. Anat. Forsch.* 21, 294.

298. MCCONKEY, E. H., and HOPKINS, J. W., 1965. 'Subribosomal particles and the transport of messenger RNA in HeLa cells': *J. Mol. Biol.* 14, 257–70.

299. MCELROY, W. D., and SELIGER, H. H., 1962. 'Biological luminescence': *Scientific American* 207, 76–89.

300. MCELROY, W. D., and SELIGER, H. H., 1962. 'Mechanism of action of firefly luciferase': *Fed. Proc.* 21, 1006–12.

301. MACGREGOR, H. C., and CALLAN, H. G., 1962. 'The actions of enzymes on lampbrush chromosomes': *Quart. J. Microscop. Sci.* 103, 173–203.

302. MCLOUGHLIN, C. B., 1961. 'The importance of mesenchymal factors in the differentiation of chick epidermis. II. Modification of epidermal differentiation by contact with different types of mesenchyme': *J. Embryol. Exper. Morph.* 9, 385–409.

303. MAGASANIK, B., 1959. 'Mechanisms for control of enzyme synthesis and enzyme activity in bacteria': in Wolstenholme, G. E. W., and O'Connor, C. M. (Eds.), *The Regulation of Cell Metabolism:* London, England: J. and A. Churchill, Ltd.

304. MANTON, I., 1952. 'Fine structure of plant cilia': *Symp. Soc. exp. Biol.* 6, 306–19.

305. MANTSAVINOS, R., and CANELLAKIS, E. S., 1959. 'Studies on the biosynthesis of deoxyribonucleic acid by soluble mammalian enzymes': *J. Biol. Chem.* 234, 628–35.

306. MARGOLIASH, E., 1963. 'Primary structure and evolution of cytochrome C': *Proc. Nat. Acad. Sci.* 50, 672–9.

307. MARKERT, C. L., 1961. 'Isozymes in kidney development': From Metcoff, J. (Ed.), *Proceedings of the Thirteenth Annual Conference on the Kidney – Hereditary developmental and immunologic aspects of kidney diseases:* Northwestern University Press, Evanston, Illinois.

308. MARSH, R. E., COREY, R. B., and PAULING, L., 1955. 'An investigation of the structure of silk fibroin': *Biochim. Biophys. Acta* 16, 1–34.

309. MARTIN, R. G., 1963. 'The first enzyme in histidine biosynthesis: the nature of feedback inhibition by histidine': *J. Biol. Chem.* 238, 257–68.

310. MARTIN, R. G., MATTHAEI, J. H., JONES, O. W., and NIRENBERG, M. W., 1960. 'Ribonucleotide composition of the genetic code': *Biochem. Biophys. Res. Comm.* 6, 410–14.

311. MATTHAEI, J. H., JONES, O. W., MARTIN, R. G., and NIRENBERG, M. W., 1962. 'Characteristics and composition of RNA coding units': *Proc. Nat. Acad. Sci.* 48, 666–77.
312. MATTHAEI, J. H., and NIRENBERG, M. W., 1961. 'Characteristics and stabilization of DNAase-sensitive protein synthesis in E. coli extracts': *Proc. Nat. Acad. Sci.* 47, 1580–602.
313. MAZIA, D., 1955. 'The organization of the mitotic apparatus': *Symp. Soc. exper. Biol.* No. 16, 355–7.
314. MAZIA, D., 1960. 'The analysis of cell reproduction': *Ann. N. Y. Acad. Sci.* 90, 455–69.
315. MAZIA, D., 1961. 'How cells divide': *Scientific American* 205, 100–20.
316. MERRIAM, R. W., 1961. 'On the fine structure and composition of the nuclear envelope': *J. Biophys. Biochem. Cytol.* 11, 559–70.
317. MESELSON, M., and STAHL, F. W., 1958. 'The replication of DNA in *Escherichia coli*': *Proc. Nat. Acad. Sci.* 44, 671–82.
318. MILLER, S. L., 1955. 'Production of some organic compounds under possible primitive earth conditions': *J. Amer. Chem. Soc.* 77, 2351–61.
319. MILLER, W. H., RATCLIFF, F., and HARTLINE, H. K., 1961. 'How cells receive stimuli': *Scientific American* 205, 222–38.
320. MIRSKY, A. E., and RIS, H., 1949. 'Variable and constant components of chromosomes': *Nature* 163, 666–7.
321. MIRSKY, A. E., and RIS, H., 1951. 'The composition and structure of isolated chromosomes': *J. Gen. Physiol.* 34, 475–92.
322. MITCHELL, P., 1963. 'Molecule group and electron translocation through natural membranes': *Biochem. Soc. Symp.* 22, 142–68.
323. MITCHELL, P., and MOYLE, J., 1958. 'Group-translocation: a consequence of enzyme-catalysed group-transfer': *Nature* 182, 372–3.
324. MONOD, J., 1958. 'Remarks on the mechanism of enzyme induction': in *Enzymes:* Academic Press, New York.
325. MONOD, J., and COHEN-BAZIRE, G., 1953. 'L'effect d'inhibition specifique dans la biosynthese de la tryptophane-desmase chez Aerobacter aerogenes': *Compt. Rend. Acad. Sci.* 236, 530–2.
326. MOODY, M., and ROBERTSON, J. D., 1960. 'The fine structure of some retinal photoreceptors': *J. Biophys. Biochem. Cytol.* 7, 87–92.
327. MORALES, M., and BOTTS, J., 1952. 'A model for the elementary process in muscle action': *Arch. Biochem. Biophys.* 37, 283–300.
328. MOREL, G., 1956. 'Nouvelles methodes permettant de realiser des cultures de tissus vegetaux': *Rev. Gen. Botan.* 63, 314–24.
329. MOSCONA, A., 1952. 'Cell suspension from organ rudiments of chick embryos': *Exper. Cell Res.* 3, 535–9.
330. MOSCONA, A., 1961. 'Rotation-mediated histogenetic aggregation of dissociated cells': *Exper. Cell Res.* 22, 455–75.
331. MOSCONA, A. A., 1961. 'How cells associate': *Scientific American* 205, 142–62.

332. Moscona, A. A., 1962. 'Analysis of cell recombinations in experimental synthesis of tissues *in vitro*': *J. Cell. Comp. Physiol.* 60, 65–80.
333. Moscona, A. A., 1963. 'Studies on cell aggregation: Demonstration of materials with selective cell-binding activity': *Proc. Nat. Acad. Sci.* 49, 742–47.
334. Moscona, M. H., and Moscona, A. A., 1966. 'Inhibition of cell aggregation *in vitro* by puromycin': *Exper. Cell Res.* 41, 703–6.
335. Moses, M. J., 1956. 'Chromosomal structures in Crayfish spermatocytes': *J. Biophys. Biochem. Cytol.* 2, 215–17.
336. Moses, M. J., 1960. 'Patterns of organization in the fine structure of chromosomes': *Proc. 4th Intern. Cong. Elec. Micros.* 2, 199–211.
337. Nagata, T., 1963. 'The molecular synchrony and sequential replication of DNA in Escherichia coli': *Proc. Nat. Acad. Sci.* 49, 551–9.
338. Nass, M. K., and Nass, S., 1963. 'Intramitochondrial fibers with DNA characteristics. I. Fixation and electron staining reactions': *J. Cell Biol.* 19, 593–612.
339. Nass, M., and Nass, M. K., 1963. 'Intramitochondrial fibers with DNA characteristics. II. Enzymatic and other hydrolytic treatments': *J. Cell Biol.* 19, 613–30.
340. Neuberger, A. (Ed.), 1958. *Symposium on Protein Structure.* John Wiley and Sons, New York, and Methuen, London.
341. Neurath, H., and Bailey, K. (Eds.), 1953–4. *The Proteins,* vols. 1–2: Academic Press, New York.
342. Nevo, A., deVries, A., and Katchalsky, A., 1955. 'Interaction of the basic polyamino acids with the red blood cells, I. Combination of polylysine with single cells': *Biochim. Biophys. Acta* 17, 536–47.
343. Newton, W. A., Beckwith, J. R., Zipser, D., and Bremner S., 1965. 'Nonsense mutants and polarity in the *Lac* operon of *Escherichia coli*': *J. Mol. Biol.* 14, 290–6.
344. Nirenberg, M., Leder, P., Bernfield, M., Brimacombe, R., Trupin, J., Rottman, F., and O'Neal, C., 1965. 'RNA codewords and protein synthesis. VII. On the general nature of the RNA code': *Proc. Nat. Acad. Sci.* 53, 1161–8.
345. Nirenberg, M. W., and Matthaei, J. H., 1961. 'The dependence of cell-free protein synthesis in *E. coli* upon naturally occurring or synthetic polyribonucleotides': *Proc. Nat. Acad. Sci.* 47, 1588–602.
346. Nishimura, S., Jacob, T. M., and Khorana, J. G., 1964. 'Synthetic deoxyribopolynucleotides as templates for ribonucleic acid polymerase: the formation and characterization of a ribopolynucleotide with a repeating trinucleotide sequence': *Proc. Nat. Acad. Sci.* 52, 1494–1501.
347. Nishimura, S., Jones, D. S., Ohtsuka, E., Hayatsu, H., Jacob, T. M., and Khorana, H. G., 1965. 'Studies on polynucleotides. XLVII. The *in vitro* synthesis of homopeptides as directed by a ribopolynucleotide containing a repeating trinucleotide sequence. New

codon sequences for lysine, glutamic acid and arginine': *J. Mol. Biol.* 13, 283–301.

348. NISHIMURA, S., JONES, D. S., and KHORANA, H. B., 1965. 'Studies on polynucleotides. XLVIII. The *in vitro* synthesis of a copolypeptide containing two amino acids in alternating sequence dependent upon a DNA-like polymer containing two nucleotides in alternating sequence': *J. Mol. Biol.* 13, 302–24.
349. NIU, M. C., 1963. 'The mode of action of ribonucleic acid': *Dev. Biol.* 7, 379–93.
350. NIU, M. C., CORDOVA, C. C., NIU, L. C., and RADBILL, C. L., 1962. 'RNA-induced biosynthesis of specific enzymes': *Proc. Nat. Acad. Sci.* 48, 1962–9.
351. NIU, M. C., and TWITTY, V., 1953. 'The differentiation of gastrula ectoderm in medium conditioned by axial mesoderm': *Proc. Nat. Acad. Sci.* 39, 985–9.
352. NOLL, H., STAEHILIN, T., and WETTSTEIN, F. O., 1963. 'Ribosomal aggregation engaged in protein synthesis: ergosome breakdown and messenger ribonucleic acid transport': *Nature* 198, 632–8.
353. NORMAN, A., and VEOMETT, R. C., 1961. 'Ribonuclease activity at the HeLa cell surface': *Virology* 14, 497–8.
354. NORTHCOTE, D. H., 1963. 'The nature of plant cell surfaces': *Biochem. Soc. Symp.* 22, 105–25.
355. NOVIKOFF, A. B., 1960. 'Biochemical and staining reactions of cytoplasmic constituents' in Rudnick, D. (Ed.), *Developing Cell Systems and Their Control:* Ronald Press, New York.
356. OPARIN, A. I., 1961. *Life: its nature, origin and development:* Oliver and Boyd, Edinburgh and London.
357. OPARIN, A. I., 1962. 'Origin and evolution of metabolism': *Comp. Biochem. Physiol.* 4, 371–7.
358. PALADE, G. E., 1953. 'An electron microscope study of the mitochondrial structure': *J. Histochem. Cytochem.* 1, 188–211.
359. PALADE, G. E., 1955. 'A small particle component of the cytoplasm': *J. Biophys. Biochem. Cytol.* 1, 59–68.
360. PALADE, G. E., 1956. 'The endoplasmic reticulum': *J. Biophys. Biochem. Cytol.* Suppl. 2, 85–97.
361. PALADE, G. E., and PORTER, K. R., 1954. 'Studies on the endoplasmic reticulum. I. – Its identification in cells *in situ*': *J. exper. Med.* 100, 641–656.
362. PARK, R. P., 1965. 'Substructure of chloroplast lamellae': *J. Cell Biol.* 27, 151–61.
363. PATTERSON, M. K., Jr., and TOUSTER, O., 1962. 'Intracellular distribution of sialic acid and its relationship to membranes': *Biochim. Biophys. Acta* 56, 626–8.

364. PAUL, J., 1962. 'Mechanisms of metabolic control in cultured mammalian cells': from Green, James W. (Ed.), *New Developments in Tissue Culture:* Rutgers University Press, New Brunswick, New Jersey.
365. PAUL, J., and FOTTRELL, P. F., 1963. 'Mechanism of D-glutamyltransferase repression in mammalian cells': *Biochim. Biophys. Acta* 67, 334–6.
366. PAUL, J., and GILMOUR, R. S., 1966. 'Restriction of deoxyribonucleic acid template activity in chromatin is organ-specific': *Nature* 210, 992–3.
367. PAUL, J., and GILMOUR, R. S., 1966. 'Template activity of DNA is restricted in chromatin': *J. Mol. Biol.* 16, 242–4.
368. PAULING, L., and COREY, R. B., 1951. 'Atomic coordinate and structure factors for two helical configurations of polypeptide chains': *Proc. Nat. Acad. Sci.* 37, 235–40.
369. PAULING, L., and COREY, R. B., 1951. 'Configurations of polypeptide chains with favored orientiations around single bonds: Two new pleated sheets': *Proc. Nat. Acad. Sci.* 37, 729–40.
370. PAULING, L., and COREY, R. B., 1952. 'Configuration of polypeptide chains with equivalent cis amide groups': *Proc. Nat. Acad. Sci.* 38, 86–93.
371. PAULING, L., COREY, R. B., and BRANSON, H. R., 1951. 'The structure of proteins: Two hydrogen-bonded helical configurations of the polypeptide chain': *Proc. Nat. Acad. Sci.* 37, 205–11.
372. PAVLOVSKAYA, T. E., and PASYNSKII, A. G., 1960. 'The original formation of amino acids under the action of ultraviolet rays and electric discharges': in Florkin, M. (Ed.), *Aspects of the Origin of Life:* Pergamon Press, London.
373. PEACOCK, W. J., 1963. 'Chromosome duplication and structure as determined by autoradiography': *Proc. Nat. Acad. Sci.* 49, 793–801.
374. PEACOCK, W. J., 1965. 'Chromosome replication': *Nat. Cancer Inst. Monogr.* 18, 101–19.
375. PEASE, D., 1955. 'Fine structure of the kidney seen by electron microscopy': *J. Histochem.* and *Cytochem.* 3, 295–308.
376. PELLING, C., 1959. 'Chromosomal synthesis of ribonucleic acid as shown by incorporation of uridine labeled with tritium': *Nature* 184, 655–6.
377. PENN, R. D., 1966. 'Ionic communication between liver cells': *J. Cell Biol.* 29, 171–4.
378. PERRY, R. P., 1960. 'On the nucleolar and nuclear dependence of cytoplasmic RNA synthesis in HeLa cells': *Exper. Cell Res.* 20, 216–20.
379. PERRY, R. P., 1962. 'The cellular sites of synthesis of ribosomal and 4S RNA': *Proc. Nat. Acad. Sci.* 48, 2179–86.
380. PERRY, R. P., 1965. 'The nucleolus and the synthesis of ribosomes': *Nat. Cancer Inst. Monogr.* 18, 325–38.
381. PERRY, R. P., HELL, A., and ERRERA, M., 1961. 'The role of the nucleolus and ribonucleic acid and protein synthesis. I. Incorporation of

cytidine into normal and nucleolar inactivated HeLa cells': *Biochim. Biophys. Acta* 49, 47–57.

382. PERRY, R. P., SRINIVASAN, P. R., and KELLY, D. E., 1964. 'Hybridization of rapidly labeled nuclear ribonucleic acids': *Science* 145, 504.
383. PERUTZ, M. F., ROSSMANN, M. G., CULLIS, A. F., MUIRHEAD, H., WILL, G., and NORTH, A. C. T., 1960. 'Structure of haemoglobin: a three-dimensional Fourier synthesis at 5·5 Å resolution, obtained by X-ray analysis': *Nature* 185, 416–22.
384. PETHICA, B. A., 1961. 'The physical chemistry of cell adhesion': *Exper. Cell Res.* Suppl. 8, 123–40.
385. PINCHOT, G. B., and HORMANSKI, M., 1962. 'Characterization of a high energy intermediate of oxidative phosphorylation': *Proc. Nat. Acad. Sci.* 48, 1970–7.
386. PLAUT, W., NASH, D., and FANNING, T., 1966. 'Ordered replication of INA in polytene chromosomes of *Drosophila melanogaster*': *J. Mol. Biol.* 16, 85–93.
387. POGO, A. O., POGO, B. G. T., LITTAU, V. C., ALLFREY, V. G., MIRSKY, A. E., and HAMILTON, M. G., 1962. 'The purification and properties of ribosomes from the thymus nucleus': *Biochim. Biophys. Acta* 55, 849–64.
388. POLLISTER, A. W., and POLLISTER, P. F., 1957. 'The structure of the Golgi apparatus': *Int. Rev. Cytol.* 6, 85–106.
389. PONTECORVO, G., and KÄFER, E., 1958. 'Genetic analysis based on mitotic recombination': *Advance Genetics* 9, 71–104.
390. PORTER, K. R., 1954. 'Electron microscopy of basophilic components of cytoplasm': *J. Histochem.* 2, 346.
391. PORTER, K. R., 1955. 'The fine structure of cells': *Fed. Proc.* 14, 673.
392. PORTER, K. R., 1956. 'The submicroscopic morphology of protoplasm': *Harvey Lect.* 51, 175.
393. POTTER, D. D., FURSHPAN, E. J., and LENNOX, E. S., 1966. 'Connections between cells of the developing squid as revealed by electrophysiological methods': *Proc. Nat. Acad. Sci.* 55, 328–36.
394. POTTER, V. R., and AUERBACH, V. H., 1959. 'Adaptive enzymes and feedback mechanisms': *Lab. Invest.* 8, 495–500.
395. PREISS, J., and BERG, P., 1960. 'Incorporation of ATP-C^{14} into polyribonucleotides': *Fed. Proc.* 19, 317.
396. PULLMAN, B., and PULLMAN, A., 1959. 'The electronic structure of the respiratory coenzymes': *Proc. Nat. Acad. Sci.* 45, 136–44.
397. REICHARD, P., 1960. 'Formation of deoxyguanosine 5′-phosphate from guanosine 5′-phosphate with enzymes from chick embryos': *Biochim. Biophys. Acta* 41, 368–9.
398. REICHARD, P., CANELLAKIS, Z. N., and CANELLAKIS, E. S., 1960. 'Regulatory mechanism in the synthesis of deoxyribonucleic acid in vitro': *Biochim. Biophys. Acta* 41, 558–9.

399. REID, C., 1957. *Excited states in chemistry and biology:* Academic Press, New York.
400. REVEL, J. P., FAWCETT, D. W., and PHILPOTT, C. W., 1963. 'Observations on mitochondrial structure. Angular configurations of the cristae': *J. Cell Biol.* 16, 187–95.
401. REVEL, J. P., ITO, S., and FAWCETT, D. W., 1958. 'Electron micrographs of myelin figures of phospholipid simulating intracellular membranes': *J. Biophys. Biochem. Cytol.* 4, 495–8.
402. RICHTER, D. (Ed.), 1957. *Metabolism of the Nervous System:* Pergamon Press, New York.
403. RIKER, A. J., and HILDEBRANDT, A. C., 1958. 'Plant tissue cultures open a botanical frontier': *Ann. Rev. Microbiol.* 12, 469–90.
404. RIS, H., 1960. 'Fine structure of the nucleus during spermiogenesis': *4th Int. Conf. Electron Micros.* 2, 199 (Springer-Verlag, Berlin, 1958).
405. RIS, H., 1961. 'Ultrastructure and molecular organization of genetic systems': *Can. J. Genet. Cytol.* 3, 95–120.
406. RIS, H., 1962. 'Ultrastructure of certain self-dependent cytoplasmic organelles': *5th Inter. Congress Electron Microscopy* 2, xx–i: Academic Press, New York.
407. RIS, H., and PLAUT, W., 1962. 'Ultrastructure of DNA-containing areas in the chloroplast of *Chlamydomonas*': *J. Cell. Biol.* 13, 383–91.
408. ROBERTSON, J. D., 1958. 'A molecular theory of cell membrane structure': *4th Int. Conf. Electron Micros.* 2, 159–171.
409. ROBERTSON, J. D., 1959. 'The ultrastructure of cell membranes and their derivatives': *Biochem. Soc. Symp.* 16, 3–43.
410. ROBERTSON, J. D., 1962. 'The membrane of the living cell': *Scientific American* 206, 64–72.
411. ROBERTSON, R. N., 1960. 'Ion transport and respiration': *Biol. Rev.* 35, 231–64.
412. ROGERS, H. J., 1963. 'The surface structures of bacteria': *Biochem. Soc. Symp.* 22, 55–100.
413. RONDABUSH, R. L., 1933. 'Phenomenon of regeneration in everted Hydra': *Biol. Bull.* 64, 253–8.
414. SAGER, R., 1960. 'Genetic systems in Chlamydomonas': *Science* 132, 1459–65.
415. SAGER, R., and ISHIDA, M. R., 1963. 'Chloroplast DNA in Chlamydomonas': *Proc. Nat. Acad. Sci.* 50, 725–30.
416. SALAS, M., SMITH, M. A., STANLEY, W. M., WAHBA, A. J., and OCHOA, S., 1965. 'Direction of reading of the genetic message': *J. Biol. Chem.* 240, 3988–95.
417. SANDBORN, E., KOEN, P. F., MCNABB, J. D., and MOORE, G., 1964. 'Cytoplasmic tubules in animal cells': *J. Ultrastr. Res.* 11, 123–38.
418. SANGER, F., 1952. 'The arrangement of amino acids in proteins': *Adv. in Protein Chem.* 7, 1–67.

419. SCHERRER, K., and DARNELL, J. E., 1962. 'Sedimentation characteristics of rapidly labelled RNA from HeLa cells': *Biochem. Biophys. Res. Comm.* 7, 486–90.
420. SCHMITT, F. O., and FESCHWIND, N., 1957. 'The axon surface': *Prog. in Biophys. & Biophys. Chem.* 8, 165–215.
421. SCHMITT, F. O., GROSS, J., and HIGHBERGER, J. H., 1955. 'States of aggregation of collagen': *Symp. Soc. Exper. Biol.* 9, 148.
422. SCHNEIDER, W. C., and KUFF, E. L., 1965. 'The isolation and some properties of rat liver mitrochondrial deoxyribonucleic acid': *Proc. Nat. Acad. Sci.* 54, 1650–8.
423. SCOTT, B. I. H., 1962. 'Electricity in plants': *Scientific American* 207, 107–17.
424. SENGEL, P., 1958. 'Determinisme de la differentiation regionale des phaneres de l'embryon de poulet': *Bull. Soc. Zool. France* 83, 82–6.
425. SHAFFER, B. M., 1957. 'Aspects of aggregation in cellular slime moulds. I. Orientation and chemotaxis': *Amer. Nat.* 91, 19–35.
426. SIBATANI, A., de KLOET, S. R., ALLFREY, V. G., and MIRSKY, A. E., 1962. 'Isolation of a nuclear RNA fraction resembling DNA in its base composition': *Proc. Nat. Acad. Sci.* 48, 471–7.
427. SIMARD-DUQUESNE, N., and COUILLARD, P., 1962. 'Amoeboid movement, I. Reactivation of glycerinated models of Amoeba proteus with adenosinetriphosphate': *Exper. Cell Res.* 28, 85–91.
428. SIMINOVITCH, L., and GRAHAM, A. F., 1956. 'The metabolic stability of nucleic acids in Escherichia coli': *Canad. J. Microbiol.* 2, 585–97.
429. SIRLIN, J., 1962. 'The Nucleolus': from *Progress in Biophysics and Biophysical Chemistry*, vol. 12, pp. 25–66: Pergamon Press, London.
430. SJÖSTRAND, F. S., 1953. 'The ultrastructure of the outer segments of rods and cones of the eye as revealed by the electron microscope': *J. Cell. Comp. Physiol.* 42, 15–44.
431. SJÖSTRAND, F. S., 1954. 'The ultrastructure of mitochondria': *8th Symp. Cell Biol., Leyden* 16.
432. SJÖSTRAND, F. S., 1956. 'The ultrastructure of cells as revealed by the electron microscope': *Int. Rev. Cytol.* 5, 455–553.
433. SJÖSTRAND, F. S., 1959. 'Fine structure of cytoplasm: The organization of membranous layers': *Rev. Modern Phys.* 31, 301.
434. SJÖSTRAND, F. S., 1962. 'The connections between A- and I-Band filaments in striated frog muscle': *J. Ultrastr. Res.* 7, 225–46.
435. SJÖSTRAND, F. S., and ANDERSSON-CEDERGREN, E., 1957. 'The ultrastructure of the skeletal muscle myofilaments at various conditions of shortening': *J. Ultrastr. Res.* 1, 74–108.
436. SJÖSTRAND, F. S., and ELFVIN, L. G., 1962. 'The layered asymmetric structure of the plasma membrane in the exocrine pancreas cell of the cat': *J. Ultrastr. Res.* 7, 504–34.
437. SJÖSTRAND, F. S., and HANZON, V., 1954. 'Membrane structure of

cytoplasm and mitochondria in exocrine cells of mouse pancreas as revealed by high resolution electron microscopy': *Exper. Cell Res.* 7, 393–414.

438. SJÖSTRAND, F. S., and HANZON, V., 1954. 'The ultrastructure of the Golgi apparatus of exocrine cells of mouse pancreas': *Exper. Cell Res.* 7, 415–29.

439. SKOOG, F., and MILLER, C. O., 1957. 'Chemical regulation of growth and organ formation in plant tissues cultured *in vitro*': in *The Biological Action of Growth Substances.* Soc. Exper. Biol., Cambridge U. Press, Cambridge.

440. SLEIGH, M. A., 1961. 'An example of mechanical co-ordination of cilia': *Nature* 191, 931–2.

441. SOLARI, A. J., 1965. 'Structure of the chromatin in sea urchin sperm': *Proc. Nat. Acad. Sci.* 53, 503–11.

442. SONNEBORN, T. M., 1959. 'Kappa and related particles in Paramecium': *Adv. in Virus Research,* 231–356.

443. SPEMANN, H., 1938. *Embryonic Development and Induction:* Yale University Press, New Haven, Conn.

444. SPENCER, M., FULLER, W., WILKINS, M. H. F., and BROWN, G. L., 1962. 'Determination of the helical configuration of ribonucleic acid molecules by X-ray diffraction study of crystalline amino-acid-transfer ribonucleic acid': *Nature* 194, 1014–20.

445. SPEYER, J. F., LENGYEL, P., BASILIO, C., and OCHOA, S., 1962. 'Synthetic polynucleotides and the amino acid code, II': *Proc. Nat. Acad. Sci.* 48, 63–8.

446. SPEYER, J. F., LENGYEL, P., BASILIO, C., and OCHOA, S., 1962. 'Synthetic polynucleotides and the amino acid code, IV'; *Proc. Nat. Acad. Sci.* 48, 441–8.

447. SPIEGEL, M., 1955. 'The reaggregation of dissociated sponge cells': *Ann. N.Y. Acad. Sci.* 60, 1056–78.

448. STADTMAN, E. R., COHEN, G. N., LEBRAS, G., and DE ROBICHON-SZULMAHSTER, H., 1961. 'Feedback inhibition and repression of aspartokinase activity in *Escherichia coli* and *Saccharomyces cerevisiae*': *J. biol. Chem.* 236, 2033–8.

449. STANLEY, W. M., and BOCK, R. M., 1961. 'Mechanisms of expression of genetic information': *Nature* 190, 299–300.

450. STEELE, R. H., and SZENT GYÖRGYI, A., 1957. 'On excitation of biological substances': *Proc. Nat. Acad. Sci.* 43, 477–91.

451. STEINBERG, M. S., 1962. 'On the mechanism of tissue reconstruction by dissociated cells, III. Free energy relations and the reorganization of fused, heteronomic tissue fragments': *Proc. Nat. Acad. Sci.* 48, 1769–76.

452. STERN, H., 1956. 'The physiology of cell division': *Ann. Rev. Plant Physiol.* 7, 91–114.

453. STERN, H., 1962. 'Function and reproduction of chromosomes': *Physiol. Revs.* 42, 271–96.
454. STEWARD, F. C., 1963. 'The control of growth in plant cells': *Scientific American* 209, 104–13.
455. STOECKENIUS, W., 1963. 'Some observations on negatively stained mitochondria': *J. Cell. Biol.* 17, 443–54.
456. STOVE, J. L., and STANIER, R. Y., 1962. 'Cellular differentiation in stalked bacteria': *Nature* 196, 1189–92.
457. STRAUS, W., 1958. 'Colorimetric analysis with N, N-dimethyl-p-phenylenediamine of the uptake of intravenously injected horseradish peroxidase by various tissues of the rat': *J. Biophys. Biochem. Cytol.* 4, 541–550.
458. SWANSON, C. P., 1957. *Cytology and Cytogenetics:* Prentice Hall, Englewood Cliffs, New Jersey.
459. SZENT-GYÖRGYI, A., 1951. *Chemistry of Muscular Contraction*, 2nd ed.: Academic Press, New York.
460. SZENT-GYÖRGYI, A., 1960. 'Submolecular biology': *Rad. Res.* Suppl. 2, 4–18.
461. TAKEUCHI, M., and TERAYAM, H., 1965. 'Preparation and chemical composition of rat liver cell membranes': *Exper. Cell Res.* 40, 32–3.
462. TARTAR, V., 1961. *The Biology of Stentor:* Pergamon Press, New York.
463. TATA, J. R. 1963. 'Inhibition of the biological action of thyroid hormones by actinomycin D and puromycin': *Nature* 197, 1167–8.
464. TAYLOR, A. C., 1966. 'Microtubules in the microspikes and cortical cytoplasm of isolated cells': *J. Cell Biol.* 28, 155–68.
465. TAYLOR, J. H., 1958. 'The duplication of chromosomes': *Scientific American* 198, 36–42.
466. TAYLOR, J. H., 1959. 'Autoradiographic studies of nucleic acids and proteins during meiosis in *Lilium longiflorum*': *Am. J. Botany* 46, 477–484.
467. TAYLOR, J. H., 1960. 'Nucleic acid synthesis in relation to the cell division cycle': *Ann. N.Y. Acad. Sci.* 96, 409.
468. TAYLOR, J. H., WOODS, P. S., and HUGHES, W. L., 1957. 'The organization and duplication of chromosomes as revealed by autoradiographic studies using tritium-labelled thymidine': *Proc. Nat. Acad. Sci.* 43, 122–8.
469. TEORELL, T., 1962. 'Excitability phenomena in artificial membranes': *Biophys. J.* 2, 27–52.
470. TERZAGHI, E., OKADA, Y., STREISINGER, G., TSUGITA, A., INOUYE, M., and EMRICH, J., 1965. 'Properties of the genetic code deduced from the study of proflavine-induced mutations': *Science* 150, 387.
471. THACH, R. E., CECERE, M. A., SUNDARAJAN, T. A., and DOTY, P., 1965. 'The polarity of messenger translation in protein synthesis': *Proc. Nat. Acad. Sci.* 54, 1167–3.

472. THOMSON, R. Y., PAUL, J., and DAVIDSON, J. N., 1958. 'The metabolic stability of the nucleic acids in cultures of a pure strain of mammalian cells': *Biochem. J.* 69, 553–61.

473. TIEDEMANN, H., 1959. 'Neue Ergebnisse zur Frage nach der chemischen Natur der Induktionsstoffe beim Organisatoreffekt Spemanns': *Die Naturwiss.* 22, 613–23.

474. TOCCHINI-VALENTINI, G. P., STODOLSKY, M., AURISICCHIO, A., SARNAT, M., GRAZIOSI, F., WEISS, S. B., and GEIDUSCHEK, E. P., 1963. 'On the asymmetry of RNA synthesis *in vivo*': *Proc. Nat. Acad. Sci.* 50, 935–54.

475. TOIVONEN, S., 1953. 'Bone-marrow of the guinea pig as a mesodermal inductor in implantation experiments with embryos of Trituris': *J. Emb. Exper. Morph.* 1, 97–104.

476. TROSKO, J. E., and WOLFF, S., 1965. 'Strandedness of *Vicia faba* chromosomes as revealed by enzyme digestion studies': *J. Cell Biol.* 26, 125–35.

477. TUNBRIDGE, R. E. (Ed.), 1957. *Connective Tissue: A.C.I.O.M.S. Symposium:* Blackwell Scientific Publications, Oxford.

478. UMBARGER, H. E., and BROWN, B., 1958. 'Isoleucine and valine metabolism in *Escherichia coli*. VII. A negative feedback mechanism controlling isoleucine biosyntheses': *J. Biol. Chem.* 233, 415–20.

479. UREY, H. C., 1952. *The Planets: Their origin and development:* Yale University Press, New Haven, Conn.

480. USSING, H. H., 1957. 'General principles and theories of membrane transport': in Murphy, R. R. (Ed.), *Metabolic Aspects of Transport across Cell Membranes:* University of Wisconsin Pres s, Madison, Wis.

481. VAN DEN TEMPEL, M., 1958. 'Distance between emulsified oil globules upon coalescence': *J. Coll. Sci.* 13, 125–33.

482. VENDRELY, R., 1955. 'The deoxyribonucleic acid content of the nucleus': in Chargaff, E., and Davidson, J. N. (Eds.), *The Nucleic Acids*, vol. 2: Academic Press, New York.

483. VILLEE, C. A., HAGERMAN, D. D., and JOEL, P. B., 1960. 'An enzymatic basis for the physiologic functions of estrogens': *Rec. Prog. Hormone Res.* 16, 46–9.

484. VOGEL, H. J., 1957. 'Repression and induction as control mechanisms of enzyme biogenesis; the "adaptive" formation of acetylornithinase': in McElroy, W. D., and Glass, B. (Eds.), *The Chemical Basis of Heredity:* Johns Hopkins Press, Baltimore.

485. VOLKIN, E., and ASTRACHAN, L., 1956. 'Intracellular distribution of labeled ribonucleic acid after phage infection of Escherichia coli': *Virology* 2, 433–7.

486. VOLKIN, E., and ASTRACHAN, L., 1957. 'RNA metabolism of T-2 infected Escherichia coli': in McElroy, W. D., and Glass, B. (Eds.), *The Clinical Basis of Heredity:* Johns Hopkins Press, Baltimore.

487. WAHBA, A. J., BASILIO, C., SPEYER, J. F., LENGYEL, P., MILLER, R. S., and OCHOA, S., 1962. 'Synthetic polynucleotides and the amino acid code VI': *Proc. Nat. Acad. Sci.* 48, 1683–6.

488. WAHBA, A. J., GARDNER, R. S., BASILIO, C., MILLER, R. S., SPEYER, J. F., and LENGYEL, P., 1963. 'Synthetic polynucleotides and the amino acid code, VIII': *Proc. Nat. Acad. Sci.* 49, 116–22.

489. WALD, G., 1955. 'The Origin of Life': in *The Physics and Chemistry of Life:* Simon and Schuster, New York.

490. WARNER, J. R., KNOFF, P. M., and RICH, A., 1963. 'A multiple ribosomal structure in protein synthesis': *Proc. Nat. Acad. Sci.* 49, 122–9.

491. WARNER, R. C., SAMUELS, H. H., ABBOTT, M. T., and KRAKOW, J. S., 1963. 'Ribonucleic acid polymerase of Azotobacter vinelandii, II: Formation of DNA-RNA hybrids with single-stranded DNA as primer'. *Proc. Nat. Acad. Sci.* 49, 533–8.

492. WATSON, J. D., and CRICK, F. H. C., 1953. 'Molecular structure of nucleic acids: a structure for deoxypentose nucleic acids': *Nature* 171, 737–8.

493. WATSON, M. L., 1955. 'The nuclear envelope. Its structure and relation to cytoplasmic membranes': *J. Biophys. Biochem. Cytol.* 1, 257–70.

494. WATSON, M. L., 1959. 'Further observations of the nuclear envelope of animal cells': *J. Biophys. Biochem. Cytol.* 6, 147–56.

495. WEBER, G., and MACDONALD, H., 1961. 'Role of enzymes in metabolic homeostasis': *Exper. Cell Res.* 22, 292–302.

496. WEBER, H. M., 1958. *The Motility of Muscles and Cells:* Harvard University Press, Cambridge.

497. WEBSTER, G., 1961. 'Protein synthesis by isolated ribosomes': in Harris, R. J. C. (Ed.), *Protein Biosynthesis:* Academic Press, New York.

498. WEIGERT, M. G., and GAREN, A., 1965. 'Base composition of nonsense codons in E. coli': *Nature* 206, 992–4.

499. WEISS, L., 1961. 'Sialic acid as a structural component of some mammalian tissue cell surfaces': *Nature* 191, 1108–9.

500. WEISS, L., 1962. 'Cell movement and cell surfaces: a working hypothesis': *J. Theoret. Biol.* 2, 236–50.

501. WEISS, P., 1945. 'Experiments on cell and axon orientation in vitro: the role of colloidal exudates in tissue organization': *J. Exper. Zool.* 100, 353–86.

502. WEISS, P., 1953. 'Some introductory remarks on the cellular aspects of differentiation': *J. Emb. Exper. Morph.* 1, 181–211.

503. WEISS, P., 1958. 'Cell contact': *Intern. Rev. Cytol.* 7, 391–422.

504. WEISS, P., 1961. 'Guiding principles in cell locomotion and cell aggregation': *Exper. Cell Res.* Suppl. 8, 260–81.

505. WEISS, S. B., 1960. 'Enzymatic incorporation of ribonucleoside triphosphates into the interpolynucleotide linkages of ribonucleic acid': *Proc. Nat. Acad. Sci.* 46, 1020–30.

506. WEISSMAN, S. M., SMELLIE, R. M. S., and PAUL, J., 1960. 'Studies on

the biosynthesis of deoxyribonucleic acid by extracts of mammalian cells. IV. The phosphorylation of thymidine': *Biochim. Biophys. Acta* 45, 101–10.

507. WHITTAM, R., 1962. 'The asymmetrical stimulation of a membrane adenosine triphosphatase in relation to active cation transport': *Biochem. J.* 84, 110–18.

508. WIENER, J., SPIRO, D., and LOWENSTEIN, W. R., 1965. 'Ultrastructure and permeability of nuclear membranes': *J. Cell Biol.* 27, 107–17.

509. WILKINS, M. H. F., and RANDALL, J. T., 1953. 'Crystallinity in sperm heads: molecular structure of nucleoprotein in vivo': *Biochim. Biophys. Acta* 10, 192–3.

510. WILKINS, M. H. F., STOKES, A. R., and WILSON, H. R., 1953. 'Molecular structure of deoxypentose nucleic acids': *Nature* 171, 738–40.

511. WILKINS, M. H. F., ZUBAY, G., and WILSON, H. R., 1959. 'X-ray diffraction studies of the molecular structure of nucleohistone and chromosomes': *J. Mol. Biol.* 1, 179–85.

512. WILLMER, E. N., 1961. 'Steroids and cell surfaces': *Biol. Rev.* 36, 368–98.

513. WISCHNITZER, S., 1958. 'An electron microscope study of the nuclear envelope of amphibian oocytes': *J. Ultrastructure Research* 1, 201–22.

514. WOLFE, S. L., 1965. 'Isolated microtubules': *J. Cell Biol.* 25, 408–13.

515. WOLFF, ET., and WOLFF, EM., 1952. 'Le determinisme de la differenciation sexuelle de la syrinx du canard cultivee in vitro': *Bull. Biol.* 86, 325–49.

516. WYNGAARDEN, J. B., and ASHTON, D. M., 1959. 'Feedback control of purine biosynthesis by purine ribonucleotides': *Nature* 183, 747–8.

517. YAMADA, T., 1962. 'The inductive phenomenon as a tool for understanding the basic mechanism of differentiation': *J. Cell. Comp. Physiol.* 60, 49–64.

518. YANAGISAWA, K., 1963. 'Genetic regulation of protein biosynthesis at the level of the ribosome?': *Biochem. Biophys. Res. Comm.* 10, 226–31.

519. YANKOFSKY, S. A., and SPIEGELMAN, S., 1962. 'The identification of the ribosomal RNA cistron by sequence complementarity, I. Specificity of complex formation': *Proc. Nat. Acad. Sci.* 48, 1069–78.

520. YANKOFSKY, S. A., and SPIEGELMAN, S., 1962. 'The identification of the ribosomal RNA cistron by sequence complementarity, II. Saturation of and competitive interaction at the RNA cistron': *Proc. Nat. Acad. Sci.* 48, 1466–72.

521. YANKOFSKY, S. A., and SPIEGELMAN, S., 1963. 'Distinct cistrons for the two ribosomal RNA components': *Proc. Nat. Acad. Sci.* 49, 538–44.

522. YANOFSKY, C., COX, E. C., and HORN, V., 1966. 'The unusual mutagenic specificity of an *E. coli* mutater gene': *Proc. Nat. Acad. Sci.* 55, 274–81.

523. YATES, R. A., and PARDEE, A. B., 1956. 'Control of pyrimidine biosynthesis in *Escherichia coli* by a feedback mechanism': *J. Biol. Chem.* 221, 757–70.

524. YATES, R. A., and PARDEE, A. B., 1957. 'Control by uracil of formation of enzymes required for orotate synthesis': *J. Biol. Chem.* 227, 677–92.

525. YCAS, M., and VINCENT, W. S., 1960. 'A ribonucleic acid fraction from yeast related in composition to desoxyribonucleic acid': *Proc. Nat. Acad. Sci.* 46, 804–11.

526. YOSHIKAWA, H., and SUEOKA, N., 1963. 'Sequential replication of Bacillus subtilis chromosome, I. Comparison of marker frequencies in exponential and stationary growth phases': *Proc. Nat. Acad. Sci.* 49, 559–66.

527. ZAMECNIK, P. C., STEPHENSON, M. L., and HECHT, L. I., 1958. 'Intermediate reactions in amino acid incorporation': *Proc. Nat. Acad. Sci.* 44, 73–8.

528. ZIEGLER, D. M., LINNANE, A. W., GREEN, D. E., DASS, C. M. S., and RIS, H., 1958. 'Studies on the electron transport system. XI. Correlation of the morphology and enzymic properties of mitochondrial and submitochondrial particles': *Biochim. Biophys. Acta* 28, 524–38.

529. ZINDER, N. D., and LEDERBERG, J., 1952. 'Genetic exchange in Salmonella': *J. Bacteriol.* 64, 679–99.

530. ZUBAY, G., and WATSON, M. R., 1959. 'The absence of histone in the bacterium *Escherichia coli*. I. Preparation and analysis of nucleoprotein extract': *J. Biophys. Biochem. Cytol.* 5, 51–4.

Index